DEVELOPMENT AND DIVERSITY

DEVELOPMENT AND DIVERSITY

New Applications in Art Therapy

EDITED BY DOUG SANDLE

Foreword By Tessa Dalley

FREE ASSOCIATION BOOKS / LONDON / NEW YORK

Published in 1998 by
Free Association Books Limited
57 Warren Street, London W1P 5PA
and 70 Washington Square South,
New York, NY 10012–1091

ISBN 1 85343 401 9 hardback; 1 85343 402 7 paperback.

A CIP catalogue record for this book is available
from the British Library.

Produced for Free Association Books Ltd by
Chase Production Services, Chadlington, OX7 3LN
Colour printing by The Witney Press
Printed in the EC by J.W. Arrowsmith, Bristol

Contents

List of Illustrations

Table

Colour Plates

The colour plates are bound near the front of the book.

Notes on Contributors

Editor

Doug Sandle BA (Hons), C.Psychol, AFBPsS is a graduate in Psychology and Sociology and has for many years worked in Higher Education, particularly with students of art, design and architecture where he has taught perceptual psychology and the psychology of aesthetics at both undergraduate and postgraduate level. Currently Reader in Visual Studies in the Faculty of Health and Environment at Leeds Metropolitan University, he has formerly held a number of Senior Academic posts, including Head of the Department of Visual Studies, and Assistant Dean in the Faculty of Cultural and Education Studies. He is the founding and continuing Director of the Leeds Art Therapy School and has been an external examiner at a number of universities, also being an adviser to the Post Graduate Arts Therapies Programme at the University of Hertfordshire. He is the founding and current Chair of the Trustees and Board of Directors of Axis, a company for database information on art and visual artists within the UK. He has been published both academically and creatively, including poetry and short stories. His radio play, *Allen Road*, has been broadcast by both BBC Radio Three and by Radio New Zealand.

Contributors

Felicity Aldridge BA(Hons), PGDip ATh has a degree in textiles and fashion, and further trained as an art therapist at Goldsmith's College, London University. She is a trained Post Adoption Counsellor, and is at present in the last year of the MA course in Art Psychotherapy at Goldsmith's. Having worked for many years as an art therapist with children for Social Services and the National Health Service, she has specialised in working with loss and trauma.

John Birtchnell MD, FRCPsych, FBPsS, DPM, Dip Psychother, RATh qualified as a doctor and trained as a psychiatrist. After six years in the National Health Service, and a brief training in psychotherapy, he became Scientific Officer with the Medical Research Council, but maintained a clinical interest in psychotherapy and marital therapy, eventually becoming honorary Senior Lecturer at the Institute of Psychiatry and honorary Consultant Psychiatrist at the Maudsley Hospital in London. He has published extensively in the psychiatric and psychological literature, was editor of the British Journal of Medical Psychology, and is the author of *How Humans Relate: A New Interpersonal Theory* (Psychological Press 1966). He is a self-taught art therapist and an Honorary member of the British Association of Art therapists. In 1993 he was Art Therapist in Residence at the Academy of Performing Arts, Perth, Australia.

Jo Bissonnet MA, BA (Hons), DipSW, ATC, RATh has a degree in fine art and trained as an art therapist at Goldsmith's college, and has an MA in social work from the University of East Anglia. She has worked for many years as an art therapist with children and adolescents in education within the National Health Service and Social Services. Currently, she works part time for Norfolk Social Services at a child and family resource centre. Her freelance work includes providing art therapy workshops, and she is involved in setting up a voluntary child bereavement support service in Norfolk, called Nelson's Journey.

Hilary Brosh Dip AD, ATC, PGDip AH, PGDip ATh, RATh worked as an art teacher and in community arts before training as an art therapist. She worked for many years in adult general psychiatry in a variety of hospital and community settings within the National Health Service and now works in private healthcare in Leeds.

Janek Dubowski PhD, BA (Hons), PGDip ATh, Cert Counselling is currently Associate Dean (Research) for the Faculty of Art and Design, University of Hertfordshire and was formerly the Director of the Arts Therapy Programme that includes postgraduate diploma and masters courses in art therapy, dramatherapy and dance movement therapy. He has published widely and is currently involved with research into the use of art therapy with autistic children.

Kathy Evans BA (Hons), PGDip ATh has a degree in fine art and further trained as an art therapist. She has recently completed research for a PhD at the University of Hertfordshire on the communication difficulties of children with autism. Parallel to this she has worked in mental health services and, during her research, in a special school for autism.

John Henzell Dip Art Ed, Mphil, RATh was born in England but spent his childhood in Australia where he trained as an artist. He worked in psychiatric hospitals and centres associated with antipsychiatry and other radical groups before teaching and researching since the 1970s. A founding member of the British Association of Art Therapists, he helped develop and run art and psychotherapy courses, and was formerly Director of the Centre for Art and Psychotherapy Studies at the University of Sheffield. He has contributed to several books on art and psychotherapy and his recent writing includes contributions to *Art and Music: Therapy and Research*, eds Gilroy and Lee (1955 Routledge), and *Art, Psychotherapy and Psychosis*, eds Killick and Schaverien, (1997 Routledge). He now writes, lectures and practices as a therapist in Sheffield.

Annie Hershkowitz BSc (Hons), BA (Hons), PGDip ATh, RATh took a degree in fine art before training in art therapy at Goldsmith's College, University of London. She worked freelance for five years, setting up art therapy groups in the community, as well as running a small private practice. She now works full time at Croydon Mental Health Services, NHS Bethlem and Maudsley Trust, in adult, child and adolescent psychiatry. She is also a practising artist, with work in private collections in Britain, France and the USA.

Jocelyne James BA (Hons), MSI, Dip CAT is a UKCP registered integrative arts psychotherapist and Sesame practitioner. She was the course leader of the Sesame Postgraduate Diploma Training in Drama Movement Therapy at the Central School of Speech and Drama, London, between 1993 and 1997, and has a wide variety of experience with different client groups in health, education and social services. Currently she is tutor at the Institute for the Arts in Therapy and Education in London, and has her own private practice for therapy supervision.

Sheila Knight BA(Hons), PGDip ATh, MA has a first degree in fine art sculpture, and a second degree in art and psychotherapy. Having worked full time as an art therapist since 1985 she has specialised in child, adolescent and family work within the public sector. In addition to her clinical work, she is currently a sessional tutor for the art therapy training programme and a workshop leader on the Spring School at the University of Sheffield. Her background as a draughtswoman in planning, water and archaeology, along with her continued involvement with sculpture, fuels her interest in houses, journeys and objects.

Marian Liebmann BA, MA, PGCE, CQSW, RATh has qualifications in teaching, social work and art therapy. She has worked in teaching and community work, as well as with offenders in a day centre and on probation. She currently works as art therapist with the Inner City Mental Health Team in Bristol, and with Mediation UK in mediation and conflict resolution. She is also visiting lecturer in art therapy at Bristol University and the City of Bath College. She is author of *Art Therapy for Groups* (Routledge 1986) and editor of *Art therapy in Practice* (1990), *Art Therapy with Offenders* (1994) and *Arts Approaches to Conflict* (1996), all published by Jessica Kingsley, as well as numerous other chapters and articles.

David Maclagan MA, ARCA, RATh read modern history at Oxford, studied painting at the Royal College of Art and trained as an art therapist at Goldsmith's College. He has published many articles on outsider art, psychological aspects of aesthetics, and art therapy. He is the author of *Creation Myths* (Thames and Hudson 1977). He is a member of the London Convivium for Archetypal Studies, leads image-making workshops in the UK and abroad, is involved in supervision, and is a practising painter.

Marijke Rutten-Saris SRKT, RATh is a LVSB registered supervisor and has worked for over twenty-five years, in both private and public practice, as an art therapist, art methodology teacher and supervisor with people of all kinds of ages and competencies. She is currently a teacher in the Department of Creative Therapy at the Hogeschool van Arnhem en Nijmegen, Netherlands, and is also a researcher at the Department of Forensic Psychiatry at the Rheinische Kliniken Bedburg-Hau. She is also currently a research student at the University of Hertfordshire. She is the founder of the Emerging Body Language (EBL) method.

Trudy Stewart PhD, MA, BSc is a qualified speech and language therapist based at St James' Hospital, Leeds. She has specialised for nearly twenty years in working with children and adults who stammer. She acts as an adviser to the Royal College of Speech and Language Therapists and to the British Stammering Association. She is also an editorial consultant for the *Journal of Fluency Disorders* and has written a number of books and papers on stammering.

Karl Tamminen BA (Hons), PGDip ATh has a degree in fine art. Having worked for many years with young people in a young offenders facility, and as an art therapist having experienced both child and general psychiatry, he has specialised in forensic work at the Humber Centre for Forensic Psychiatry, where he is now Head of the Art Therapy Department. He is a visiting lecturer to various professional training centres and has a keen interest in education and learning. He is currently studying for an MSc in Professional Practice in Art Therapy at the University of Lincolnshire and Humberside.

Louis Thomas BA(Hons), PGDip ATh, RATh graduated as a graphic designer, and is a qualified art therapist pursuing further training in child psychotherapy with the United Bristol Healthcare Trust. He has worked extensively with children and families, from mothers with newborn babies and adolescents who are parents themselves. For five years he worked for West Glamorgan Social Services as an art therapist, providing long-term individual work for children where sexual abuse had been a feature in their lives, as well as providing consultancy to other professionals. He currently works as a Senior Art Therapist for Gwent Community Health Trust.

Foreword

It seems to me that the purpose of any conference is twofold. The most obvious is to create a professional forum for sharing ideas and exchanging views and for delegates to take away with them their individual experiences and thoughts. The second is to find some way to mark the moment, to record the experience, and document the ideas. Conference proceedings or publications such as this enable the ideas generated at the time to reach a wider audience. The possibility of reading contributions after the event, especially when further developed, provides a reflective, more considered opportunity to absorb and digest and to take some of the thinking forward.

This book came from the conference held at Leeds Metropolitan University in May 1996, and as its title The Uses and Applications of Art Therapy suggests, the theme of the conference was essentially about practice. The contributions, with widely differing approaches, working contexts and client groups show the diversity, and also the developments within the profession as a whole. Art therapists are employed by health, social services, education and other statutory and voluntary organisations. Broad representation at the conference from these various working situations was indeed stimulating and interesting. There was a freshness and spontaneity about a genuine need to share information and think together about practice. With the growing need for accountability and professional effectiveness, such opportunities for thinking and learning from each other are to be welcomed.

Each contributor offers a particular approach. Within this broad spectrum of practice, however, the central consideration is the artwork made in the session. The image forms the foundation of all art therapy practice. How to think about the image that is made in art therapy, how this affects the relationship between client and therapist and the consideration of how the art therapist constellates around the image, creates the beginnings of divergence through different theoretical perspectives. Whatever the wider contexts of the therapy room, it becomes clear from these contributions that whether there is a blank sheet of paper, or a child's spontaneous expression, a cartoon or comic strip or a carefully considered aesthetic piece, the image remains central to the work of art therapists. However broad the range of clients, from those who stammer or have been sexually abused, children with learning difficulties and no verbal language, offenders in a secure unit or probation clients, as groups or as individuals – the list in this collection is rich and interesting – in all cases the image provides the focus for communication and thereby for understanding growth and change. The image or visual statement made in the therapy room is the starting point, the inner representation, the record of an emotional experience, the creative moment where the unconscious can be made conscious. When the non-verbal can become verbal, communication begins, marking a moment of change or transformation.

Within the UK, art therapy as a profession has recently undergone major developments with State Registration. The practising art therapist is having to adapt to the challenge of a demanding and changing world in the work place. It is at this interface of outer and inner world that the potential and possibilities of the various approaches of the art therapist can contribute to an understanding of the image and the detailed processes of its making.

Tessa Dalley
December 1997

Plate 1
A great white bird hovers over desiccated black trees.

Plate 2
A pure white phallus stands up against a tangle of hair, flesh and blood.

Plate 3
Black Trunk.

Plate 4
Peering through dark wall of deprivation.

Plate 5
‘They are both me – good and bad.’

Plate 6
Masks produced by group of boys.

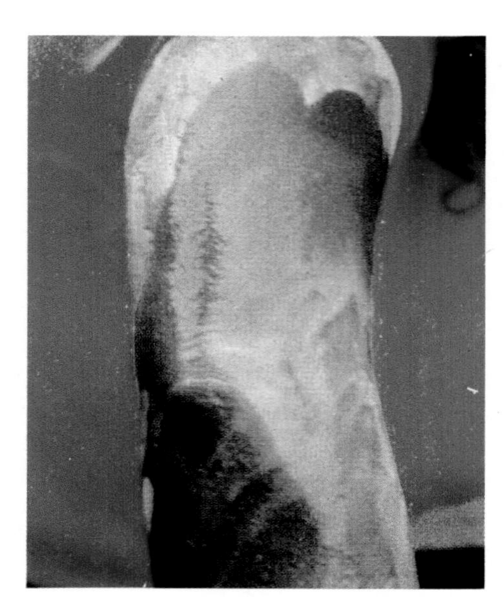

Plate 7
Arm covered in slimy green paint.

Plate 8
G's painting which helped bring back
childhood memories.

Introduction

DOUG SANDLE

In 1979 the first Leeds Art Therapy School was organised by a group of academics and practising art therapists. This project, which at the time of writing is still continuing at Leeds Metropolitan University, arose out of a then perceived need to introduce art therapy to wider, but appropriate, constituencies. Its five-day format with both core workshops and optional introductory activities has been used as a model for the establishment of similar short-course Schools, both within the United Kingdom and abroad. As then, the School attracts a diverse range of participants, and while some are established art therapists, the majority are working, or aspire to work, in a broad range of health, social care, and educational contexts.

At the time there was some caution about 'spreading the word' about art therapy too widely or too indiscriminately, and the profession, undergoing its own formative problems and issues within the UK, had other priorities. These included a concern for the maintenance of good practice and ethics, the development of appropriate structures for professional training and education, and for recognition as a valid professional practice with appropriate status and remuneration.[1] However, in aspiring for professional status and recognition there is always the danger of developing an inward-looking exclusivity. The problem is to ensure on the one hand, that professional standards, training, good models of practice and ethics are established and maintained, but on the other to ensure that the purpose and social value of art therapy is effectively propagated and developed in an accessible and socially relevant way. Thus it was felt by the organisers of the first Leeds Art Therapy School, that if art therapy had an important role to play in many diverse areas of individual and social life, then it was important that other cognate professions and concerns should have the opportunity to become familiar with art therapy praxis. From such connections it was hoped that art therapy would be welcomed as a partner profession in a range of contexts beyond, what was then, the more traditional area of psychiatry and mental health.

Inevitably, as a growing profession becomes more established and confident, its concerns will be less inward, and relationships with other cognate professional areas will develop. Today art therapists not only practice in many diverse areas but there is a growing reciprocal interest in art therapy.[2] Thus in 1995, when the organisers of the Leeds Art Therapy School hosted a conference on the Uses and Applications of Art Therapy, well over half its audience were non-art therapists, but were representative of other cognate professions and concerns. As the initial aim of the conference was to introduce different areas of practice to a wider professional constituency this was gratifying. The papers from that conference, in many cases further reworked and developed, are published in this book and represent some of the diverse working contexts and developments in art therapy practice.

The value of interdisciplinary work and collaboration between two professions is demonstrated in the opening chapter. It highlights the benefits of art therapy as a communicative process and its potential use with clients with speech impairment, such as stammering. Stewart, a speech and language therapist, and Brosh, an art therapist, present a review of the literature in which artwork is used with adults who stammer as a diagnostic, diversionary or expressive tool.

In presenting their own work, the authors illustrate the use of drawing by speech impaired clients to externalise and articulate self-perceptions and feelings about their impediment. Drawings are also used to provide dynamic insights into patient perception of the on-going speech therapy process. Stewart and Brosh present case studies in which artwork provided an invaluable adjunct to the identification phase of therapy, and in revealing some of the emotional and attitudinal components of stammer. Art is also shown to be a particularly sensitive and insightful indicator of the unfolding speech therapeutic process, and thus can be used to clarify, monitor and articulate change and development occurring within speech therapy.

Child and sexual abuse are areas where the appropriateness of art therapy has become increasingly recognised,[3] and in their introduction to their book on working with children in art therapy, Case and Dalley comment:

> The experience of art therapists working with child abuse for example, is that it is far more likely that the abuse will be expressed through 'making a mess' and an understanding can be reached through the transference and countertransference process, rather than through a drawing depicting who did what to whom, on a particular occasion. The art therapist is trained to pick up communications of great sensitivity through the process of image making, and more importantly

> waiting with, holding and containing the anxiety and uncertainty of the child struggling with the unfolding of their deepest difficulties. Therefore most art therapists hesitate to leap into early interpretation and judgement as to the meaning of the communication because of the very delicacy of the process. This would suggest that the art therapist may need to work to a point where the child can verbalise the experience with the support of the therapist through the ensuing months rather than use the actual material obtained from one particular session. (Case and Dalley 1990:4)

The contribution by Thomas is an ideal example of this process as he develops a case example of a four-year-old girl, in which the reader is able to follow the infant's struggle to communicate violent sexual abuse, and to use her art therapist to help her make sense of an overwhelming traumatic experience. This is achieved through a process in which pre-verbal visual images are gradually and sensitively understood to become dynamic representations of trauma. Particularly poignant in this account is the careful way in which the dynamic relationship between therapist and infant is established.

Thomas recognises that, even at this early age, infant mark-making has both form and function, and he shows how the process of mark-making, especially by the physical and bodily aspect involved, can in itself be communicative and revelatory. This process of physical engagement with the art material and mark-making, even at a precursory stage of image-making, becomes the focus for the therapeutic interaction, rather than just the produced image as symbol or metaphor. Through this dynamic process the child begins to inwardly acknowledge trauma, and outwardly communicates to the therapist. This process of conveying otherwise unspeakable trauma through the concrete and physical process of the artwork and its creation is developed by Thomas as 're-presentation'.

That infant mark-making can have meaning, even in apparently pre-representational forms, is also argued by Dubowski and James in their contribution on the role of art therapy, and arts therapies generally, within the context of affective education. Two case studies are presented which demonstrate the role of both visual art therapy and aspects of drama, movement and music therapy in facilitating emotional development within children with learning disabilities. The graphic narrative illustrates how the skilled therapist encourages affective articulation in a way which facilitates emotional release, self-knowledge and social interaction.

The importance of affective education is particularly urgent with children with special learning difficulties, and the role of art is again seen as pivotal to this. In particular, the symbolic and cognitive aspects of mark- and image-making also have a role in developing a learning ability for the attainment of communication skills. This is exemplified within a particular case study of art therapy with a boy with severe learning disabilities and limited use of language. The art therapy process is outlined as one which encourages the development of pictorial representation from basic undifferentiated play to an attained skill, whereby even pre-pictorial mark-making gains some significance in developing concepts of representational, and hence potentially communicable, meaning. This, it is suggested, facilitates a more general improvement in self-identity, concentration span and general behaviour. The implications for the role of art in affective education in enabling emotional development and emotional literacy are further discussed.

As in the previous two contributions, that by Evans and Rutten-Saris acknowledge the possibilities of meaningful communication with children, who initially are without developed skills of pictorial symbolisation. Their contribution reflects their mutual interest in working with autism and the application of art therapy to develop communication and levels of cognitive interaction between therapist and autistic child. They consider that their work with autism requires an alternative model to the more conventional objects-relations approach, where severe communication problems with the patient mitigates against verbal discussion around the form and content of the art imagery.

Alternatively, they apply the theories of Daniel Stern and the concept of Vitality Affects as a basis for developing an expressive relationship with their autistic subjects. In the absence of conventional communication, the expressive nuances of gesture, bodily rhythm and a Gestalt totality of cross-modal perceived qualitative behaviour provide the cues for expressive interaction between therapist and child. The therapist responds actively to such bodily cues, expressively perceiving them as Categorical Affects or mood feelings, and is able to enter into a sympathetic active relationship with the child through sharing in and shaping the creative process. In this manner, communication is developed, and the basis for a perceptual cognitive understanding is reached as a significant behavioural response within autism.

The process is illustrated by case studies which show how the therapist utilises the Vitality Affects contained in the creative process and the emerging qualities of the creative activity, rather than just the finished artwork. The Vitality Affects are encouraged by the active participation of the therapist in the creative process through sharing and developing subtle expressive interaction through an Emerging Body Language, developed by Rutten-Saris as a methodological paradigm. Examples demonstrate how the therapist engages in attuning and shaping expressive qualities as shared and mutually understood experience, which can form the basis for a more developed communication.

Important in Evans' and Rutten-Saris' contribution is the notion that the expressive nuances of marks and images and the expressive process of their production are important conveyors of meaning, and hence there is an implicit recognition of aesthetic qualities in contributing to this process. In his contribution, Maclagan recognises the aesthetic as an important aspect in the artwork of his anorectic patient, particularly the concept of the sublime, which Maclagan argues has a particular importance in a consideration of the various dimensions of experience that face, and in a sense contribute to, anorexia. These other dimensions include the cultural, anthropological, religious and political, and a notion of the psyche that includes that of soul, as developed from Hillman's notion of the soul as a dimension of lived experience.

It is this concept of soul, rather than traditional physical or psychological models that rely on physical-based metaphors, that provides a way into a notion of anorexia that acknowledges its concerns for the experiential edges that lie between pre-puberty and adult sexual identities, between maternal and patriarchal domains, between autogenesis and intersubjectivity and between life and death. Furthermore these concerns are seen in their imaginal aspect, in which anorexia is both an image in itself and one that relates to prevailing cultural images of fashion and media.

Art can embody this imaginable dimension of anorexia, but not just as content, but by stylistic characteristics (neglected by conventional psychoanalysis in favour of content) and by aesthetic concepts such as the sublime. These ideas are articulated with a consideration of the aesthetic import of the visual work produced by his patient.

In his contribution, Tamminen discusses his role as an art therapist within a secure forensic unit, and outlines the particular needs and issues that inform his practice. In particular, the need to establish 'informed consent' is pertinent in working with forensic clients. In this, the processes of art therapy, whereby common terms of reference and shared meanings need to be established, is seen as particularly apposite. Moreover this informed consent is not just about participation in art therapy, but concerns a degree of acceptance of the wish to return to society. In this, Tamminen sees the role of art therapy as providing the means for new understandings and perceptions of self and society, and of identity and expectations.

The therapeutic process begins with the criminal offence, which is seen as the end of a journey; a life journey which is backtracked to explore how self-identity has developed, and how moral codes were formed, changed and lost. Examples of imagery from a case study, in which the narrative of a painful and traumatic life journey is visually and dramatically expressed, explore how self-perception, identity and expectations impact upon criminal behaviour and the potential for change.

Liebmann focuses her work with probation clients around their offending behaviour, and then widens this to look at work with other offenders in such situations as prisons and secure units. Her early work with offenders on probation was carried out within the context of a newly emerging Criminal Justice Act, and the need was to focus on therapy that did more than help offenders in a general way, but which addressed offending behaviour directly. In reviewing the uses of art therapy with offenders, she describes how she developed the comic strip format and an art therapy process that is uniquely channelled into a particular visual format. While this might be seen to inhibit expression, Liebmann demonstrates the versatility and effectiveness of this method, not only in facilitating the co-operation of the client, but in providing important narrative clues to the nature of the offending behaviour and in developing the probationer's understanding of this. It also provides a basis for later less structured art therapy work

Three case studies are presented: a sex offender, a violent offender and a victim of sexual abuse. The comic strip technique is seen to provide a rapid means of rapport and communication as a vehicle that readily enables the probationer to develop beyond the narrative of the particular offending incident to an awareness of broader life narratives and their psychodynamics. Finally, Liebmann discusses the role of art therapy in other offender contexts.

As art therapy is more widely applied, it is important for practitioners to be able to reflect on the particular issues located within different contexts. In her contribution, Bissonnet not only outlines examples of short-term art therapy groups within a social services context, but usefully discusses some of the issues in the use of co-facilitators from other disciplines, and issues of gender within the working therapeutic relationship.

The process of how the groups were formed provides a useful model for co-operation among different disciplines, and for achieving suitable referrals in a way that ensures full knowledge and agreement by all relevant parties concerned, including parents. Important consideration is given to the relevance of art therapy both to groups and to male adolescents in particular.

The detailed examples illustrate how the group art therapy employed can be both a healing activity in itself, and a significant part of the therapeutic dynamics. Particular insights are suggested around sexuality and gender confusion among abused adolescent boys, the symbolic expressive acknowledgement of hurt, and the development of group dynamics.

Art therapy is playing an increasing role in both individual and social trauma and the contribution by Aldridge outlines the role of images and pictures in dealing with trauma, and their efficacy as a means of communication for those traumatised. While this therapeutic process has been applied with individual children, Aldridge particularly outlines its use in family therapy, and presents an example in which a family attempts to come to terms with the trauma experienced by two young brothers in witnessing a murder. Their subsequent identification of the murderer and the realisation that they might be called to give witness lead to psychological disturbance which has a dynamic effect on family relationships.

Highlighting the use of art therapy in trauma, family therapy and within a forensic context, Aldridge demonstrates the dynamic role of imagery for both the individual and for the family in coping with trauma and in providing support.

The Uses and Applications of Art Therapy conference, in being concerned principally with communicating practice, presented an opportunity for Birtchnell to present a particular way of working that he has developed as a regular workshop leader at the Annual Leeds Art Therapy School. Although placed within a group setting, the therapy sessions focus, in turn, on individuals in intense and highly concentrated sessions. Particular to the method employed by Birtchnell is the use of art as a means of working jointly with the client, where not only images, but written words are used as dynamically evolving clues to self-revelation and as a means to find and release emotion. Utilising principles adapted from Gestalt therapy, images and words are externalised within the artwork by both the client and the therapist. This intervention by the therapist within the image-making, including the use of written words, enables the artwork to become the dynamic and evolving locus for the therapeutic session. In this the artwork acts as a communicative tool to facilitate self-revelation and emotion, with aesthetic considerations regarded as of secondary importance. Birtchnell describes in detail this particular method and graphically describes the kind of outcomes achieved. The chapter ends with a particular case study where the relevance of his technique for family and interpersonal problems is suggested.

In her contribution, Knight stresses the importance of attachment, separation and identity as recurrent themes in therapeutic work with children. While this is a common concern of much psychotherapy, Knight's work centres upon attachment as physically mediated by the skin and the body. While attachment, security and belonging are all components of identity, it is the skin, the outer shell of the physical body, which becomes a vehicle for the symbolic enactment of attachment and separation. This is further transformed through clothes which act as a symbolic skin.

The actual or symbolic damaging of the skin can give clues to inner conflicts and emotions, and Knight notes how the children she has worked with often request plasters or bandages in their play, which, used to indicate damage to the skin, signify inner emotional wounding. The skin as a container of the self becomes vulnerable where there is inner turmoil, and she notes the symbolic importance of common indicators of inner distress such as self-cutting, bruising, eczema, soiling, wetting and the ripping and cutting of clothes, all involving the skin as mediator between inner and outer self.

Two case examples are presented, where through art therapy the particular use of materials as symbolic skin are seen as important clues to emotional problems. The properties of plaster and of paint, sharing properties with skin such as being laid onto surfaces, covering or protecting or cutting into as when carved or etched, provide a means to work through art with attachment problems by way of dynamic reference to skin as the border between inner and outer self.

As resources within the United Kingdom's National Health Service, and for psychiatric treatment generally, become increasingly constrained, therapy over brief but intense periods is becoming more commonplace. Art therapy, and its potential for triadic transference processes among the client, therapist and art object, is regarded by Hershkowitz as particularly suitable for focused and directed periods of short-term therapy. Working with psychiatric patients, a number of case studies are presented in which, as well as the actual subject imagery, the type of materials used and the processes they give rise to, appear to be revelatory and therapeutic. The material nature of clay and three-dimensional work for example, are regarded as particularly suitable for brief work and as having the potential to relate more to narcissistic and pre-differentiation dynamics.[4]

The final contribution, although not so directly concerned with art therapy practice as such, reminds us of the power and influence of images within individual and cultural perceptions. Henzell's contribution stems from therapeutic support given to members of the Sheffield Nine O'clock Service (NOS), as one of a team of therapists and counsellors following its collapse. Henzell is concerned with the dynamic potency of particular internalised images and their role in both the attraction for, and subsequent withdrawal from, a cult organisation. He argues that the psychodynamic import of these images and their accompanying manipulative processes, within the context of a religion that is image-bound, could exercise psychological power over anyone, not just certain susceptible personalities.

Henzell looks at the role of the priest-leader of NOS in manipulating images and imagery, often by seeking out the unconscious needs of his followers, and analyses how the ulterior intentions of these powerful and intoxicating images concealed themselves until everything unravelled in the end. Henzell argues that a particular sensitivity is required to understand the role of such imagery, one that avoids obvious therapeutic techniques or a literal response to what is taking place. Such is required to humanise and demystify the attraction and pain of the images in order to gain an immunity against past trauma, as well as against the possibility of a similar psychodynamic group phenomenon.

With the emphasis on practice, the chapters presented here outline different ways of working and different contexts of professional engagement. Although the articulation of theoretical issues was not a main concern, clearly, different theoretical perspectives underpin the contributions, and these are derived from several theoretical perspectives such as object relations, psychoanalysis, Gestalt theory, and from developmental and cognitive psychology. While there is a need for the continued development and refinement of major theoretical perspectives and the articulation of theoretical issues,[5] a characteristic of contemporary art therapy practice is its eclectic nature, or rather the process of drawing upon different theoretical perspectives to suit different contexts.[6] Similarly, the methodologies employed are often conceived to meet the needs and exigencies of particular client groups, and while, for example, Liebmann's use of cartoon drawings can apply to other client groups than offenders, and the methodology developed by Rutten-Saris can be applied to adults as well as children, contemporary practice does not seem obsessed with claims for the monolithic universality of a particular method and its application.

One of the issues touched upon during discussion at the conference was that of research and the issue of the efficacy of art therapy. The implication was that in extending the constituency and sphere of influence of art therapy, there was a need to justify its practice by research. In the following chapters, some of the contributors refer to the outcomes of their case histories as they perceive them, while others briefly articulate some of the issues around measures or evidence of efficacy.[7] Henzell, Maclagan, and Tamminen, for example, note the inappropriateness for art therapy of conventional empirical research or assessment methodologies that are derived from the natural sciences. Again, it was clear from conference discussions, that a wide spectrum of research, determined by the particular context of the therapeutic practice, which might or might not include appropriate positivistic or hermeneutic methodologies, could be employed as determined by the questions being asked, the nature of the client–therapist relationship, and the broader sociopolitical context of the practice.

In a similar respect, discussion highlighted the need for flexible and sensitive perspectives that recognised issues around gender and cultural diversity. Perhaps, given the nature of art therapy, concerned with essential human issues, the potentially problematic issue of power relationships between therapist and client, and its location within areas of social as well as personal discourse, the need for awareness and articulation of gender and race issues is increasingly recognised.[8]

Diversity of practice, of theoretical perspectives, of research methodologies and a growing awareness and accommodation of difference are all characteristic of postmodernity. They in themselves acknowledge the fragmentation and non-unitary nature of both experience and our construction, interpretation and representation of it. That a conference on art therapy practice should fail to sustain a particular theoretical perspective or an all-embracing methodology is not then surprising. Such should also be regarded as a strength, if art therapy is to extend its constituency and sphere of influence in a society and culture that is not in itself monolithic or unitary. Perhaps it is not without significance that the 1998 conference of one professional art therapy association should be entitled 'Diversity and Strength'.[9]

Notes

1 Diane Waller (1991) provides an authoritative account of the development of art therapy as a profession within the United Kingdom.

2 Many agencies, both within the private and public sector and covering a wide area of welfare, rehabilitation and therapeutic concerns, continue to be keen to sponsor participants to the Leeds Art Therapy Summer School. In 1990 the British Association of Art Therapists (1990) in its survey of the conditions of service of registered art therapists indicated that while just over half worked for the National Health Service, there were the beginnings of employment opportunities in other areas, particularly the Social Services. A more recent survey in the UK for BAAT carried out by the trades union, MSF (Carpenter 1997), asked what clinical directorates respondents had input into, and found that the main categories described were adult (71%), elderly (25%), child and adolescent (20%), learning disability (19%), rehabilitation (13.7%), forensics (10.9%), substance mis-

use (10.4%). Other specific categories mentioned overall by nearly 9% of replies included, for example, HIV and Aids, eating disorder, mental health, palliative care, continuing care, schools, community care, cancer oncology and alcohol abuse. Most respondents worked in more than one area with some in up to eight. As an example from elsewhere of a continuing growth of specialist and diverse interests within art therapy, the website of the Canadian Association of Art Therapists specifically lists sections, among others, for art therapy and learning disabilities, cancer, the treatment of trauma, Aids and HIV, and bereavement.

3 In 1995 for example, the international journal, *The Arts in Psychotherapy* (Landy 1995) devoted a special edition to sexual abuse.

4 The significant differences between two- and three-dimensional media and their implications for use in art therapy, and the particular dynamic qualities of clay and plasticine in working with schizophrenics are developed in detail by Foster (1997).

5 For example, Evans and Rutten-Saris in their chapter as well as highlighting the work of Stern, raise by implication issues related to the organismic-developmental theory of language and expression, (as also do Dubowski and Evans, and Thomas). The theoretical work of Werner and Kaplan (1963) for example, would seem an appropriate area for development in relation to many aspects of art therapy to complement the more acknowledged and growing psychotherapeutic influences within art therapy.

The value and importance of continuing theoretical development is exemplified by Schaverien (1987; 1992), whose further articulation of transference and countertransference is representative of contemporary theoretical contributions to art therapy practice. Another example is Maclagan who in this book and elsewhere, for example (1995), has demonstrated the relevance of theoretical issues within aesthetics to art therapy.

6 Wood (1997: 171) illustrates the wide diversity in differing models of practice as gleaned from books by art therapists.

While different art therapists might be consistent in their developed methodologies, eclectic adaptation using different influences for different problems is also practised. For example from interviewing a sample of art therapists drawn from those registered with BAAT and working in the three Northern areas of England, Dillen (1996) noted that none of the sample of art therapists interviewed worked very specifically within one theoretical perspective, and that they acknowledged several influences. She concluded that there was no one dominant theory of art therapy and quoted one therapist as stating 'you kind of draw in different theories all the time'.

7 With regard to psychotherapy generally, Seligman (1995) in reviewing the Consumer Reports survey, perhaps the most extensive survey undertaken on the value or otherwise of psychotherapy, makes the distinction between an efficacy study and an effectiveness study. He usefully highlights the many methodological issues involved in such research. Discussions around the appropriateness of different research traditions and models, and an articulation of the many theoretical issues involved in art therapy research are extensively articulated by several contributors in *Art and Music: Therapy and Research*, edited by Gilroy and Lee (1995). Politsky draws out the issues around positivist and hermeneutic research

paradigms for the creative arts therapies, and in discussing the implications for research in the creative therapies, summarises the postmodern character of art in contrast to the positivism of modernism as follows:

> ... postmodernism values the impurity that comes with focusing on the social, economic and political aspects of art rather than the formal. In asserting the social construction of reality, art is no longer pure. In contrast to modernism, postmodernism is tolerant of the impure, ambiguity, contradiction, complexity, incoherence and inclusiveness. Subjective and intimate, it blurs the boundaries between the world, the self and disciplines. It recycles the past, sythesizes rather than analyzes. Finally, it is psychological, narrational, filled with personal content and subjective facts. (Politsky 1995:312)

8 The growth of Women's Therapy Centres and Well Women Clinics in the UK has contributed to this process. Case (1990), for example, reports on work done at the Women's Therapy Centre in London in her consideration of feminist issues around a particular case study and Schaverien (1995) provides a contemporary theoretical perspective on gender issues in art therapy. Dalley (1990) for example, recognised the importance of racial and multi-cultural issues in working with children, and identified some of the key issues. The continuing and growing development of awareness of and sensitivity to racial issues is more recently exemplified by a special issue of *The Arts in Psychotherapy* on multiculturalism (Lewis 1997).
9 The 1998 Annual Conference of the Canadian Art Therapy Association is entitled Diversity and Strength: The Many Faces of Canadian Art Therapy. CATA runs one of the more developed websites on art therapy; (http://home.ican.net/phansen/pages/BKStrauma.html). It can also be reached via another useful and interesting website, Nick Totten's Non Mainstream Psychotherapy and Counselling Resources on the Internet at (http://ourworld.compuserve.com/homepages/selfheal/nonmain.htm).

References

British Association of Art Therapists (1990) *Survey of Conditions of Employment*.

Carpenter, S. (1997) *BAAT/MSF Survey of Art Therapists*.

Case, C. (1990) Reflections and Shadows: An Exploration of the World of the Rejected Girl, in Case, C. and Dalley, T. (1990) *Working With Children in Art Therapy*, London and New York: Routledge.

Dalley, T. (1990) Images and Integration: Art Therapy in a Multi-cultural School, in Case, C. and Dalley, T. (1990) *Working With Children in Art Therapy*, London and New York: Routledge.

Dillen, C. (1996) An Exploratory Study into the Use and 'Effectiveness' of Art Therapy with Patients who are Experiencing Behavioural Problems, unpublished project, part fulfilment of the requirements for the degree of B.Sc.(Hons) Health Studies, Leeds Metropolitan University, England.

Foster, F. (1997) Fear of Three Dimensionality: Clay and plasticine as Experimental Bodies, in Killick, K. and Schaverien, J. (1997) *Art, Psychotherapy and Psychosis*, London and New York: Routledge.

Gilroy, A and Lee, C. (1995) *Art and Music: Therapy and Research*, London and New York: Routledge.

Landy, R.J. (1995) (ed.) Special Edition on Sexual Abuse, *The Arts in Psychotherapy 22* (5).

Lewis, P. (1997) (ed.) Special Edition on Multiculturalism, *The Arts in Psychotherapy 24* (2).

Maclagan, D. (1995) The Biter Bit : Subjective Features of Research in Art and Therapy, in Gilroy, A. and Lee, C. (1995) *Art and Music: Therapy and Research*, London and New York: Routledge.

Politsky, R.H. (1995) Toward a Typology of Research in the Creative Arts Therapies, *The Arts in Psychotherapy 22* (4) 307–14.

Schaverien, J. (1987) The Scapegoat and the Talisman: Transference in Art Therapy, in Dalley, T., Case, C., Schaverien, J. et al. (1987) *Images of Art Therapy: New developments in theory and practice*, London and New York: Routledge.

—— (1992) *The Revealing Image: Analytic Art Psychotherapy in Theory and Practice*, London and New York: Routledge.

—— (1995) *Desire and the Female Therapist: Engendered Gazes in Psychotherapy and Art Therapy*, London and New York: Routledge.

Seligman, M.E.P. (1995) The Effectiveness of Psychotherapy: The Consumer Reports Study, *American Psychologist 50* (12) 965–74.

Waller, D. (1991) *Becoming a Profession: The History of Art Therapy In Britain 1940–82*. London and New York: Tavistock/Routledge.

Werner, H. and Kaplan, B. (1963) *Symbol Formation: An Organismic-Developmental Approach to Language and the Expression of Thought*, New York, London and Sydney: John Wiley & Sons.

Wood, C. (1997) The History of Art Therapy and Psychosis 1938–95, in Killick, K. and Schaverien, J. (1997) *Art, Psychotherapy and Psychosis*, London and New York: Routledge.

1 The Use of Drawings in the Management of Adults Who Stammer[1]

TRUDY STEWART and HILARY BROSH

This chapter arose from a collaboration between an art therapist and a speech and language therapist, specialising in work with non-fluent clients. The particular application of art therapy concerns the use of clients' own drawings in the management of adults who stammer. While the benefits of art therapy are well documented, its particular use with this client group is minimal. Two case studies are presented that illustrate several practical advantages of this approach.

One of the concerns of art therapy is with the use of art materials for the expression of thoughts and feelings that otherwise may be difficult to describe in words. As such, it can offer an alternative form of self-expression and communication of powerful or confusing emotions such as anger, hate or love. Issues surrounding change and loss, previously taboo subjects such as sexuality or death, and negative self-perceptions can be safely expressed through a variety of art materials and safely contained in an image within a therapeutic relationship.

Art therapists work with many different types of clients and, using a humanistic and holistic approach, stammering clients can be managed in the same way as any other client group. In such an application, it would be inappropriate to identify one single meaning in an image for diagnostic purposes. The uniqueness of individual artistic expression and the many layers of meaning contained in a single image should be recognised. Similarly, attention should be given to the process of image-making as well as to the outcome. This means acknowledging the feelings, thoughts and memories that may emerge from the self-reflection that is part of the creative process, as well as the physical and tactile responses to the media, the techniques used and the sometimes surprising images that appear on the paper.

As the client takes responsibility and control for what is portrayed, it is the client who is empowered to set the agenda, and a picture can emerge of concerns and feelings that are different from what he or she reports verbally. Liebmann (1990), in describing the many advantages of using art therapy, refers to the cathartic effect possible with the use of art as a safe and acceptable way of dealing with unacceptable emotions. Wadeson (1980) similarly discusses the possibility of using a form of self-expression that does not have to obey the rules of sequential logic or grammatical structures.

Case and Dalley (1992), writing about a young client using art therapy to work through unresolved bereavement issues, highlight the importance of giving time in an art therapy session for responding to and reflecting on the images produced. This reflection may mean verbalising feelings that have emerged in the image-making process; sometimes as crying or laughing. More often, it involves sitting in silence in front of an image that exists as a tangible, non-verbal container and transmitter of feelings. These feelings can be 'read' from the image and experienced within the transference and counter-transference phenomena of a therapeutic relationship. Thus, non-verbal communication has always been seen as an important part of the process. In Dalley, Naumberg, one of the post-war pioneers of art therapy in America, is cited as describing the expression of emotions and experiences in images as being less threatening than verbal expression:

> The techniques of art therapy are based on the knowledge that every individual, whether trained or untrained in art, has a latent capacity to project his inner conflicts into visual form. As patients picture such inner experiences it frequently happens that they become more verbally articulate. (Dalley 1984: 13)

Naumberg's statement takes on an added dimension in the context of working with individuals with communication difficulties, such as adults who stammer. One of the earliest research papers that used illustrations by such clients was by Wischner (1952). In an attempt to prove the hypothesis that stammering behaviour is rewarded through the reduction of anxiety, thirty-three stammerers were asked to draw whatever they thought most adequately represented their behaviour immediately before, during, and after a moment of stuttering. The resulting drawings were analysed in terms of the content of the illustrations, the spontaneous written language that accompanied the drawings and the process or cycle of events that were represented. Wischner concluded that the content and language illustrated progressively mounting tension, anxiety or stress followed by a reduction in this state upon completion of the stuttering act, which seemed to confirm the original hypothesis. Ten years later, Wischner's study was replicated by Sheehan, Cortese and Hadley (1962) with forty-eight subjects and with greater rigour (i.e. control of experimenter bias, observer reliability and other variables). Sheehan noted marked individual differences in the stammering behaviour as represented in the drawings. He concluded that stammering may be rewarded in a variety of ways, and the nature of the reinforcement could also be different for different stammerers. Bar and Jakab (1968), a few years later, suggested the use of drawings as a projective technique to identify stammerers' underlying feelings and motivations. In addition they suggested that drawings could be used as a diagnostic tool to guide the planning of therapy.

A review of more recent literature reveals infrequent application of visualisation in the general management of adults who stammer. Overall, there appears to be a limited number of functions to which art therapy has been applied, as follows:

1. *Diagnostic*: Devore, Nandur and Manning (1984) used drawings as a measure of communication attitudes. Similarly Rinsky (1994) interpreted and evaluated speech disorders from drawings.

2. *Diversionary*: Two papers from the Soviet Union described a use of art therapy as a corrective or 'diversionary' technique. Kharatinov (1981) freed clients from what he called a 'fixation on their stuttering' and Kasparov and Miranova (1983) described the use of drawing instruction to correct stammering in the developing vocabulary of pre-school stammerers.

3. *Expressive*: Two papers presented at the First World Congress on Fluency Disorders in Munich in 1994 described the use of drawing as an expressive, creative activity. Schmidt (1994) used a method that combined different art therapies to correct stammering. It was stated that creativity and self-expression transform the stuttering children's inner tension into a positive form. Eikeland (1994) described drawing as an alternative channel of communication for emotional expression. Both of these papers used drawing in a similar way to the art therapy approach described above.

Of particular interest was the work by Cheasman, Logan and Moffatt at the City Lit. Institute, London, who used art therapy within group therapy for adults who stammer (Anderson 1994). As part of the identification phase of therapy, clients were asked to create images to represent their feelings about stammering now and as a child, to see how these feelings might have changed. They report varying reactions to the task, but believe that the images reveal deeper feelings than the spoken word.

For a number of years, therapists in Leeds working with adults who stammer have used clients' drawings in a very limited way in therapy. For example, clients were asked to draw their stammer quite early on in the therapy process, often as part of an identification phase. Little direction was given regarding the nature of the illustration, the form it might take or the medium that could be used (i.e. pen and ink, coloured pencils and crayons or felt tip pens). Clients were encouraged to be creative and to generate their own ideas.

In general, two types of illustrations emerged. Firstly, there were those clients who represented a speaking situation that aroused extreme emotion; for example, queuing in a shop or speaking in front of a group of people. Often what emerged from these drawings was the client's perception of being the centre of attention, frequently before he or she spoke. These drawings also facilitated discussion of the key areas perceived by the client, i.e. the face of the bank clerk, and the feelings evoked by the figure immediately behind him or her in the queue.

Other clients illustrated their speaking problems in a different way, conveying similar depth of emotion in an abstract form. One client felt his stammer, which had held him back over the years, was a constant drain on his energy and saw it as a great burden he shouldered. Thus, he drew a figure struggling up a steep flight of stairs called life, with the burden of stammering weighing him down, while others apparently forged ahead of him with ease (Figure 1.1).

Figure 1.1 The stammerer's burden.

While working with this client group, it became apparent that for some people the use of drawing in therapy was an illuminating experience. They related well to the medium and it appeared to enable them to construe their difficulties in a different way. It was not clear why some clients responded so positively to this opportunity and why others did not. It may have been due to their previous experiences and use of drawing and their perceived abilities or lack of abilities with the medium. There may also have been a more fundamental inhibition within some clients that had previously prevented them from 'letting go'.

Nevertheless, the two case studies that follow demonstrate the usefulness of this approach. Each demonstrates how drawings were used in therapy, albeit in quite different ways, to further the understanding of the clients' problems from their own perspective and to facilitate the process of change.

In the first example, A, a fifty-year-old graphic designer, was married with one teenage son. He had attended a group therapy programme in 1978–79 that had taught a 'speak more fluently' approach. He referred himself for therapy in 1994, as he was experiencing significant problems with fluency and had high levels of both word and situational avoidance.

Assessment revealed a mild overt stammer, characterised by sound, syllable and word repetitions with associated diaphragmatic, clavicular, laryngeal and facial tension. Occasional severe dysfluencies occurred on plosives (for example t, k), with loss of eye contact and backward-jerking head movements. More significant covert symptoms of word, situational and self-role avoidance were noted. He also reported high levels of anxiety related to a number of specific speaking situations.

A was offered a place on a non-intensive group therapy programme that was significantly different from his previous group therapy sessions, both in terms of philosophy and structure. (The recent sessions were more interactive in style with the general aim of clients being able to say what they wanted to say when they wished to say it. The teaching of fluency enhancing techniques was included in some sessions as required, in response to individual client's needs rather than to the group as a whole.)

Following a number of sessions in which A experimented with both non-speech and speech change, therapy focused on identifying the features of A's stammer. To supplement the understanding of his overt stammering behaviour, A was asked to draw his stammer. This he did at home and the resulting picture was brought to a therapy session for display and group discussion.

The illustration (Figure 1.2) was surprising from the clinician's perspective for a number of reasons. Firstly, the anatomical detail

Figure 1.2 A's self portrait of his stammer.

appeared to be so accurate that it was possible to locate specific muscle groups and areas of tension associated with A's dysfluency from the drawing alone. Secondly, it yielded several new areas of information that previously had not come to light, despite the fact that A had attended therapy a number of years previously and the clinicians considered themselves to be familiar with his speech patterns. Finally, the drawing produced an emotional response in the clinicians who saw it. One felt it conveyed pain and anguish and likened the image to that of a torture victim. Another clinician, who also felt there was pain in the picture, noted a starkness and vulnerability and commented that the stammer had been laid bare.

In the therapy session, A was asked to talk about his picture to the other members of the group. A number of interesting points emerged regarding how he perceived and felt about his stammer, as follows:

1. *Muscle detail.* A was able to identify specific muscle groups involved in his stammer. He discussed in particular, muscles down from the jaw to the neck, at the base of the skull, the back of the neck and in his face at the side. Interestingly, he described frequently pulling neck muscles during blocks, and of being aware of joints or vertebrae in his neck creaking after a series of prolonged hard blocks or bouts of stammering. He was also aware of the involvement of non-speech muscles, such as those around his shoulders and in his neck.

2. *Increased pressure.* The drawing highlighted this feature present in his head. He reported feeling this pressure increase along with a rise in temperature, especially around his temples, during prolonged stammering. Raised pressure was also noted in his chest. He described being 'full to capacity' and yet trying to inhale further during a stammer. The film drawn across his mouth indicated his lack of ability to exhale and/or make sound.

3. *Perceiving others.* The illustration clearly showed A's attitude and perceptions of other people. He was aware of his own lack of eye contact, as shown in the unseeing, empty eye sockets, but conscious of others being present around him. They took on less significance in relation to the figure himself and A described his own focus as on his stammering.

The second example, B, was twenty-six years old at the time of his referral and was studying with a view to starting a university course the following year. He was the youngest child in a family of four where the parents had had a turbulent relationship. This had culminated in divorce when he was eight years old and he had subsequently lived with his mother, providing nursing care during her terminal illness. She died when he was fourteen.

B was involved in a serious road traffic accident in 1991, which resulted in severe head and facial injuries. He made a remarkable recovery, although still suffering from occasional bouts of dizziness since the accident. He had two sessions of speech and language therapy at nine years of age but none since that time. At the time of his referral, he was living at home with his father, brother and an elderly relative.

On initial assessment he presented with a moderate stammer characterised by long silent blocks, with associated jaw jerks, hard articulatory attack and loss of eye contact. Other overt symptoms included eye blinking, nasal flaring and increased muscular tension, which was observed and specifically reported in his legs, shoulders and diaphragm. Avoidance at word level was not reported as a covert symptom, although B did identify a number of other problems. He discussed avoidance of situations (e.g. telephoning), rehearsing speech and anxiety related to speaking situations.

B was offered individual speech and language therapy sessions in the first instance. Within these sessions, a Personal Construct approach was used to explore a number of personal and family issues that seemed to have affected B's construing of himself and his speech. In addition, Anxiety Control Training was taught (Snaith 1981; Turnbull 1987) and work on identification of both overt and covert aspects of his stammer began. After four sessions of individual therapy, it was suggested that B attend group therapy sessions for adults who stammer. These sessions were to run in parallel with the individual sessions for a time.

It was at this point that B was also asked to draw his stammer. It had been noted in the individual sessions that he used a lot of visual images when discussing his difficulties (e.g. animal metaphors: 'I feel like a tortoise. I would like to be a lion.') and the clinician considered this a possible indicator of the usefulness of a visual medium. Again, no specific instructions were given regarding the form or content of the illustration. However, reassurance was provided on the acceptability of any drawing and on the non-judgemental approach that would be used by the clinician.

B carried out the task at home and he took time to complete it. No pressure was exerted on him to finish the drawing by the therapist. Finally, B brought the drawing to an individual session and it was discussed both then and in a subsequent group session (Figure 1.3).

Figure 1.3 B's drawing of his stammer.

On first seeing the illustration, the clinician felt overwhelmed by the content. The depth of emotion, the isolation and sense of sadness were intense. During the session, the clinician and client spent several moments in silence looking at the picture. It was noted by the clinician that B was a talented illustrator and the prediction that this medium might be useful had been demonstrated beyond expectations. Later in the session, the clinician chose to clarify a number of specific features and details from the drawing. The client commented on the profound dichotomy of the two scenes; each one being the opposite of the other. He described how this was also reflected in the way he had constructed the illustration, drawing each scene on its own sheet and then gluing them together. Reflecting on this he said: 'Perhaps what is represented here are two opposing mentalities; the one I felt I was stuck with and the one I hoped to find' (Birdsall 1995).

In ensuing discussion, it was noted that there were no other people on the island. B commented that they were all inside enjoying themselves. B was also asked why he had chosen to include three buildings and whether this had some significance. The question evoked a strong reaction from B as he realised, for the first time, that the buildings represented his family. Further discussion resulted from this revelation, including confrontation of the isolation and separateness he experienced from his family. In both group and individual therapy, strategies to enable him to reach the island were put forward, including swimming, waiting for the tide and attracting other people's attention. None of these suggestions was met with a positive response. Thus, the image seemed to be further compounded by a sense that B was self-absorbed and focused on his current situation to the exclusion of possible ways forward. (This view was at odds with the desire for change, which he vehemently stated in individual therapy sessions.)

Four months after the original drawing, B was asked to repeat the exercise. Again possible medium and content were not discussed, although there was an understanding that this second illustration would have some relation to the first drawing, and might reveal the changes he had been able to make in both individual and group therapy.

In the intervening therapy sessions, considerable progress had been made both in terms of attitude and specific speech behaviour changes. These can be summarised as follows:

1. He had tackled some major issues regarding perceived family attitudes to himself and his speech.

2. He had confronted the loss of significant relationships in his life, including the death of his mother and the break-up of an

important, long-standing relationship following his road traffic accident in 1991.

3. He had worked on a more open attitude to stammering, including routinely mentioning his speech in everyday conversations with a variety of people, discussing stammering with peers and family members, maintaining eye contact during dysfluent speech and voluntary stammering.

4. He experimented with both non-speech and speech change.

5. He had worked on developing a more assertive approach rather than an aggressive response, which had been more his usual pattern.

6. He identified features of his speech that were contributing to his dysfluency: clavicular tension, irregular breathing pattern, facial tension.

7. He reduced specific aspects of his stammering, notably 'back tracking', avoidance at several levels, and began work on block modification.

Throughout therapy, references had been made to the original drawing, both specific detail and the emotions that it portrayed. It was thought that a second drawing at this stage of therapy would allow further reflection for the client on the progress and changes he had made in relation to his starting point (i.e. the raft), and to where he wanted to be (i.e. the island). The clinician also considered that a second drawing would provide a type of subjective assessment of progress; a creative visual representation of an outcome measure using a self-rating by the client.

This illustration was again drawn at home and brought to and discussed in a subsequent therapy session. There were a number of points arising out of this second drawing (Figure 1.4), which formed the basis of discussion with the clinician in individual therapy. Firstly, there was the obvious difference and change in the two drawings. Clearly B had moved considerably from his original perception of himself and indeed, had moved in the direction he wished to go. There appeared to be significant differences in both how he saw himself and where he wished to move to. The figure in the second drawing was upright, stable and had a strength about him. His grasp on 'the key' was firm.

Figure 1.4 B's second drawing, showing a change in self-perception.

Secondly, it was clear from the illustration that B now believed he had a choice. This was in marked contrast to his previous drawing where there appeared to be no way out and no alternative to his predicament. In this later image, he seemed to be weighing up his options and there was a sense of uncertainty about the future. It could be that the doubt had arisen from his closer view of 'the island'. Having moved to a position where he faced the possibility of achieving what he had set out to achieve, the reality looked slightly different. Similarly, doubt could have arisen from the very fact that he was now able to make a choice, which may have been a new, and threatening experience for him.

In discussions that took place in clinic, B himself commented on the changes depicted in the drawing. Once again, the illustrations allowed him to reflect on where he was in therapy and how he felt:

> I drew this picture at a time when I personally felt I was on the precipice of a fundamental change in my life. I think I had come to a

> crossroads and a choice of route was needed. I could continue along that tried and tested path of 'B the stammerer' which, much as I disliked it, was predictable: the devil I knew. On the other hand I could take this new path, which offered me all the things I had long hoped for. However, walking that path and getting those things also meant taking my armour off, leaving myself open, taking risks. I think this picture captures that uncertainty. (Birdsall 1995)

B was discharged from both group and individual therapy one year after his initial interview. He left the area and began a university degree course. At this point, the clinician contacted him and asked him to consider whether he would like to draw a final image. He agreed to think about it. Some weeks later a third drawing was received. A further clinical session was arranged to discuss the drawing and B's comments were recorded.

The clinician awaited the third picture with anticipation (Figure 1.5). Initially it evoked a sense of pleasure and enjoyment as everyone depicted seemed to be having a good time. The assumption was made that B was one of the people in the drawing, but it was not immediately clear which one he was. This was interpreted as good integration. On a more detailed analysis, the naked figure was identified and this raised questions about B's vulnerability and integration.

The evidence of further change was also immediately apparent. It appeared that he had arrived at his desired destination and seemed relatively comfortable with his environment. In terms of communication, he had illustrated himself in a speaking situation, which was

Figure 1.5 Drawing by B after discharge.

interpreted as a major development. The depiction of the weather was also thought to be significant. Both cloud and rain appeared in equal amounts, which contrasted markedly with the previous drawings. This seemed to indicate a more balanced, realistic view of life in which both good and bad times might occur in the future.

In discussion with B, issues of vulnerability and openness were considered. He commented that, while he had a greater self-perception, he had limited understanding of other people; hence the faces of others were featureless. His focus on himself and his stammer had led to the exclusion of others, which he readily acknowledged at this stage.

Whereas in the previous drawings, there had been some indication of where B was aiming to reach (i.e. the island, one of the buildings), in this latest picture there was no such indication, and so the clinician asked how he viewed his future. B commented:

> To be quite honest, I do not think he will go anywhere else. This is about as good as it gets. If I was asked to draw the next episode of this I do not think I could. It would only be about more of the same. At this stage of my life I do not feel a need to go anywhere. It's okay here. (Birdsall 1995)

As our earlier review of the art therapy literature showed quite clearly, art therapists are involved frequently with clients who have communication problems, usually in the context of other psychiatric or psychological difficulties. However, there were no examples of art therapists working with the emotional aspects of stammering. Similarly, a review of the fluency literature in Europe and America showed that the specific use of drawing is rarely used with stammering clients and even less in Britain with adults who stammer. The greater incidence of use with children may reflect a view that drawing is a childlike activity that has little or no relevance to other age groups. If, however, the benefits of this approach are to be more widely appreciated, then there needs to be a fundamental change in the attitude to drawing so that it is not regarded solely as a diversionary activity, nor as the prerogative of younger clients and those who work with them.

However, there are a number of documented benefits of this approach. Liebmann (1990) lists a number of advantages, with particular reference to work with offenders, which are reported by her elsewhere in this book.

The detailed case studies that have been presented in this chapter particularly illustrate the benefits of using drawings with adults who stammer. One case demonstrated how a client's illustration was an invaluable adjunct to the identification phase of therapy. Initially presented as a possible means of recognising overt symptoms, it proved useful in identifying some emotional and attitudinal components of the stammer. The second case study showed how drawings were used at various points in therapy in clarification, monitoring, consolidation and as tangible evidence of the client's stages of change. In this instance it allowed the client to reflect, in a concrete and tangible way, on a number of important issues. These included his stammer, the stages of change he experienced and the outcome he wished to achieve.

Of particular interest to speech and language therapists is the possibility of using this approach as a qualitative outcome measure. Its potential for work with adult clients has yet to be fully explored, but may prove a useful adjunct to other self-rating scales, especially if used collaboratively between two professions, such as art therapy and speech and language therapy.

Note

1 With some minor changes within the present context this contribution is reprinted with permission from the *Journal of Fluency Disorders 22* (1) Stewart, T. and Brosh, H., The Use of Drawings in the Management of Adults who Stammer, 35–50, 1997, © Elsevier Science Inc.

References

Anderson, J. (1994) New Directions at the City. *Signal 1* (1).

Bar, A & Jakab, I. (1968) Graphic Identification of the Stuttering Episode as Experienced by Stutterers. *Psychiatry and Art 2* 2–15.

Birdsall, M. 1995. Changing The Way I Speak: Is It What I Do Or What I Think? Presentation at the British Stammering Association Conference.

Case, C. and Dalley, T. (1992) *The Handbook of Art Therapy*. London: Tavistock/Routledge.

Dalley, T. (1984) *Art as Therapy*. London: Tavistock.

Devore, J.E., Nandur, M.S. & Manning, W.H. (1984) Projective Drawings and Children Who Stutter. *Journal of Fluency Disorders 9* (3) 217–26.

Eikeland, K. (1994) Drawings and Psychodrama. Alternative Channels of Communicating with Stutterers: A Therapist's Considerations. *Journal of Fluency Disorders 19* (3) 217–26.

Kasparov, V.I. & Miranova, S. (1983) Use of Decorative Drawing Classes for Stuttering Correction in Pre-schoolers. *Nursery School-Kindergarten 993* (3) 58–60.

Kharatinov, A.V. (1981) Methodological Principles of Art Therapy in a System of Social Rehabilitation for Adult Stutterers. *Novye Issledovaniya v Psikhologii 1* (24) 94–7.

Liebmann, M. (1990) *Art Therapy in Practice*. London: Jessica Kingsley.

Rinsky, K. (1994) The Examination of Stuttering Children Through Drawings. *Journal of Fluency Disorders 19* (3) 204.

Schmidt, E. (1994) Correction of Stuttering by Complex Art Therapy. *Journal of Fluency Disorders 19* (3) 206.

Sheehan, J.G., Cortese, P.A. & Hadley R.G. (1962) Guilt, Shame and Tension in Graphic Projections of Stuttering. *Journal of Speech and Hearing Disorders 27* (2) 129–39.

Snaith, R.P. (1981) *Clinical Neurosis.* Oxford: Oxford University Press.

Stewart, T. and Brosh, H. (1997) The Use of Drawings in the Management of Adults Who Stammer, *Journal of Fluency Disorders 22* (1) 35–50.

Turnbull, J. (1987) Anxiety Control Training and its Place in Stuttering Therapy in *Stuttering Therapies: Practical Approaches.* C. Levy (ed.), London: Croom Helm.

Wadeson, H. (1980) *Art Psychotherapy.* London: Wiley & Sons.

Wischner, G.J. (1952) Anxiety-reduction as Reinforcement in Maladaptive Behavior: Evidence in Stutterers' Representations of the Moment of Difficulty. *Journal of Abnormal Social Psychology, 47* (566–71).

2 From Re-Presentations to Representations of Sexual Abuse

LOUIS THOMAS

> In my opinion what grips us so powerfully can only be the artist's intention in so far as he succeeds in expressing it in his work and in getting us to understand it. I realise that this cannot be merely a matter of intellectual comprehension; what he aims at is to awaken in us the same emotional attitude, the same mental constellation which produces in him the impetus to create. (Freud 1914: 212)

In this chapter I am going to share some thoughts, feelings and information regarding my work, using a case example to illustrate some key features on the subject of child sexual abuse and its clinical treatment. The case I am going to share is not one taken from my present place of work, but instead is the very first client I took on after qualifying. (Names, identities and background details have been changed whenever necessary to protect confidentiality.) The case lends itself well towards helping describe a particular aspect of art-making called the 're-presentation' that I define as: 'the concrete evocation of trauma through art-making'; recognised when the artwork embodies (Freud 1910: 203; Pater 1873: 117; Schaverien 1987: 78) the feelings of abuse in a concrete form, rather than expressing its content symbolically or metaphorically. This is quite distinct from an artistic representation that instead stands-for or symbolises an event. The re-presentation has both a developmental and a destructive potential within the therapeutic relationship; and which route it takes depends very much upon the art therapist's response to the raw material being offered for containment.

Children are representing things all the time when they are in spontaneous play; and in this sense, if all goes relatively well, many children are and most adults have been 'artists'. It is a now familiar notion that one of the functions of play, art, and creative gesture is to represent certain fragments of our life that we find a need to psychically reorganise, so that they can be used in the service of self-development. This recognition is perhaps largely due to the work of Winnicott (1958) who, through detailed observations of children and parents in play, gave equal importance to its function for the child as preparation for forming human relationships, and as helping reduce excessive innate drive. Observing children in free play can teach us much about these processes, and if we are fortunate we can observe our own children symbolically playing out life events that need constant rehearsing to gain the understanding, confidence and survival skills necessary to ensure personality growth.

For example, we may observe the exquisite way in which a child, barely two, uniquely discovers something of the nature of her own capacity for independence when unsuccessfully attempting to harness the companionship and control of a busy earthworm on her first spring walk in the garden. Or we can witness the way even a baby, of several weeks, copes momentarily with prolonged absence by recreating presence through mouthing his or her fingers.

The young child who has experienced sexual trauma however, has a far greater complex of difficulties to solve when calling on natural creative resources for organising her experiences of independence, control and self in relation to others. In the following case example of a four-year-old girl, we follow her struggle to communicate violent sexual abuse and to use her art therapist to help her make sense of an experience that not only left her bereft of words, but was so overwhelming that she could not at first even think about the abusive scenes that were making her inner and outer worlds so unstable. The destabilising effects of abuse, disrupted therapy, investigation and her removal from home are expressed poignantly by B (as we shall call her), in her fruitless attempts to make her world more stable by compulsively sellotaping herself to large immovable pieces of furniture.

What is discovered is that, even though at first this young child was unable to think, talk or conventionally draw her frightening experiences, she became more able to communicate her trauma through a 'primitive' form of pre-art. This was achieved through a particular aspect of scribbling and mark-making, in which the pitch, gradient and tonality, to both the art object and the attuning therapist, enabled the child's needs urgently to evacuate the troubling emotional affect.

Exposure to sexual trauma can be so unthinkable for the child that she may often feel overwhelmed by her own spontaneous attempts even to represent it symbolically through art and play. Creative activity can then become a frightening experience, which for the child is extremely difficult to differentiate from the actual abuse that it is failing to symbolise. Segal (1957) saw this phenomenon as an area of 'symbolic equation', where there can be a part or total fusion between a physical event and the subject's attempts to reconstruct it psychically. Applied within the context of child sexual abuse work, this occurs when in play, the precipitating trauma and its emotional affect burst through the child's immature and damaged coping mechanisms, and she simply cannot endure the experience long enough to make any sense of it. It is just too confusing and painful. It needs to be got rid of. Yet, we now know that if the experience is not thought about and understood, it can sometimes remain in its raw state scarcely below the surface of the child's persona. And because of the creative need to express, transform and integrate it, it resurfaces for further processing; sometimes with damaging results; for example, in using other children or people in an effort to make sense of the abuse by attempting to play it with vulnerable or abusive others.

The art therapist can help intervene in the sometimes cyclical nature of sexual abuse by providing an opportunity for the confusing 'equative' feelings (Segal) to find a graphic form. These re-presentations (Thomas 1992) provide the very foundation upon which child and art therapist begin the, at times, terrible task of thinking, talking and drawing the ill-repressed experiences that can fuel distressing or dangerous behaviour. The art therapy environment becomes, if you like, a surrogate container for the unprocessed abusive recollections until the child can begin herself to emotionally contain them. By first concretely re-presenting the abuse with her art therapist the hostile experience is shared, becomes more bearable for the child and therefore more subject to her symbolising processes and 'thinking' mind. This also helps create a sense of difference between the actuality of the trauma and having thoughts about it. In effect, the child can learn to symbolise the experience and hopefully find less of a need to communicate it through action. This process is fraught with complexity and unique to each individual child, and frequently demands a long-term therapeutic relationship based on a quality that is sensitive to issues surrounding the misuse of power in intimacy.

Through first feeling that her scribbles and 'mess'-making are warmly received by another as potentially meaningful, B begins to develop the confidence and inner resources to think about the fears that emotional closeness to a man raises for her. This is in order to gain a greater perspective on what she has been through, and how it sometimes shapes her present understanding of intimacy, i.e. that 'characterised by a close or warm personal relationship' (*Oxford English Dictionary* 1989). As Fairbairn (1971: 116) states, 'The basic neurotic conflict is between dependence and independence; when the person one turns to is the person one must get away from.'

B was referred to me by her mother and her nursery supervisor, who were concerned with the levels of aggression she displayed and the general disruption this caused at home and amongst her peers. She was very frequently getting into physical fights, seemingly purposely urinating in play areas (she was toilet trained), eating distasteful materials and forever washing and hurting herself. Self-harm took on a variety of forms, but the most prominent included forearm biting and determinedly seeking sharp objects to cut at her hair. I worked with her over a period of six months, offering her three hourly sessions per week. Art therapy was conducted in a designated sluice area attached to her nursery, which although maintaining a sense of privacy, was situated within the building in such a way that she could keep other people in sight if she wished.

B's first re-presentational drawings, produced in the first session, were full of seemingly formless intense scribbles, lines and shading, their creation invested with an intense aggression that spilled out of the picture frame in the form of screams and sporadic attacks directed towards me. These aggression-laden drawings, which failed to contain what they embodied, and B's reaction towards my presence during this process, helped me to understand something of the violence inside her, which perhaps she herself had known in her external world. There were times when I would feel stuck and helpless by her attacks and her enraged demands for me to 'Stay There!', having these words yelled at me even though I sat quietly with no intention of leaving her side.

The quality of B's demands for me to 'Stay There' seemed double-edged, conveying both a fear of intrusion and a desperate wish for me to keep her company. In the art therapy relationship, B now projected these overwhelming feelings from herself into me as a defence against acute emotion, but a defence that if utilised with the wrong person, at an inopportune moment, could elicit an abusive reaction. In art therapy, B's more profound emotions were beginning to be held in the environment provided. One of the aims of this was that she and I could then make time to begin the work of creatively processing such emotions with our thoughts. Bion (1962) was very interested in the concept of 'therapist as container'. He believed part of the function of therapy was to allow the child to project into the therapist those emotions that she is too immature to have developed any means to process herself. The child requires a person who is capable of receiving, digesting and handing back the now modified projections in such a way as not to be made over-anxious or overwhelmed by them.

During these earlier stages of art therapy, helping B process her emotions involved me in both staying with her as she struggled to find the courage to express the chaotic feelings and their unknown effects upon our relationship, and being able to survive her frequent frightening attacks. Perhaps a sign that we were both beginning to contain and process some of the malignant material was when, on one occasion after several weeks of getting to know her, she approached me just before a session and offered me a doll's plastic cup near full of her urine. I received B's 'present of pee' as symbolic of her growing trust in me to accept a part of herself, perhaps experienced previously as uncontainable.

In a later session B expressed the 'dirty' feelings she had about herself and that appeared to her to colour the way she experienced many things. This particular drawing (Figure 2.1) is of 'dirty shoes; dirty laces', and B's 'dirty mouth'. She communicated and re-presented a scenario, probably at the root of her 'dirty mouth' feelings, by

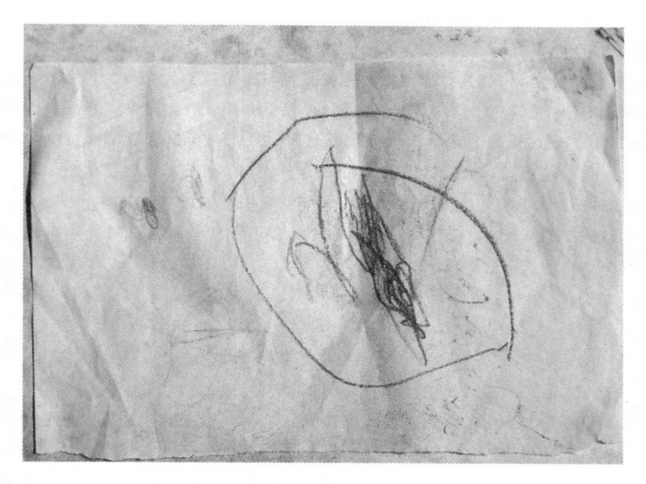

Figure 2.1 'Dirty shoes, dirty laces', and B's 'dirty mouth'.

rhythmically pushing a large navy blue crayon in escalating aggression in and out of her mouth, sucking, biting and reaching to vomit. My own gut response to the oral intrusion being played out was one of feeling quite nauseous. B attempted to gain control over the crayon, which had seemed to take on a malignant presence of its own (a not uncommon feature of re-presentational art-making), by biting off small pieces, until, when her mouth overflowed, she would call out for 'tissue' and attempt to contain the spill, vigorously wiping the inside of her mouth 'clean'. The intensity and urgency with which B laboured at these activities was noticeably greater than the kinds of messy play and oral predicaments that children of her age can sometimes find themselves in when playing unsupervised with certain art materials. There was also a marked obsessional quality to it.

Looking a little closer at her production of this image, we can consider if her artwork begins by being truly symbolic. She seems here to start by creating buffering symbols of some distressing thoughts embodied in her 'dirty shoes' and 'dirty laces' drawings. But when her mind locates 'dirty mouth' activity, (whatever this may mean to her), it all proves too much and her artwork is seen to fall in on itself, leaving her at the undiluted mercy of its unprocessed contents. When this happens, the only thing standing between the child and the potential to be re-traumatised by the image is the person of the art therapist; the enduring presence of another struggling to think about the uncontained, unthinkable contents that frighten the child. Attempts are then made, as best one can, to feed back an emotional tonality of the experience presented, transformed by the art therapist's thoughtful use of words that aim to absorb some of the inherent shock.

Perhaps the line that B has drawn around this composite symbolic/re-presentational image suggests a greater sense of emotional containment than her earlier work, coinciding with her improving capacity to trust me to be a receptacle for her frightening experiences.

In the same session, B requested painting. This, I felt, was an important step for her. It signified a growing capacity to trust me to support her further through the expression of emotion and/or events by using a medium that is more difficult to control than crayon drawing; and as such, tends to allow issues of uncontrol and power to be focused more sharply. During the creation of this image, B projected onto and into me feelings of personal intrusion, and the expectation that I would be intrusive by ordering me to cover her in paint. Painting her own arms, B would shout and scream: 'Do this. Go on.' By thinking aloud upon its content and partially digesting the emotional impact of this fiercely projected request, I attempted to keep one foot firmly in the external reality of the situation, whilst also struggling to acknowledge B's internalised state and the anxiety it created between us. An example of my voiced thoughts in such moments would be along the lines of, 'Sometimes B, it seems hard for you to *always know* that I will not mess you' (spoken with appropriate sensitivity). My speech also served to try and digest affect and reduce anxiety by communication and reinforcing a willingness on my part to share in her personal exploration, without wishing to intrude upon her self-rights. These 'aloud thoughts', when the timing was right, might also have provided B with an example of benign emotional penetration and strengthened her hope of becoming more able to develop the means to overcome this 'occasional' disability.

The first marks made by B in the painting were of a figurative quality in that they strongly suggested the image of a person or persons. Then, using copious amounts of finger paint, she smeared the figure or figures into a blobby blot and tried to claw her way through what she called the 'shit' and 'mess'. It seems that B was communicating to me the wish to be helped in finding a way to bear what she was at pains to be hidden from. It may be that for B, the need to bear 'pain' and the need to forget pain, were indistinguishable.

Figure 2.2 'This is you'.

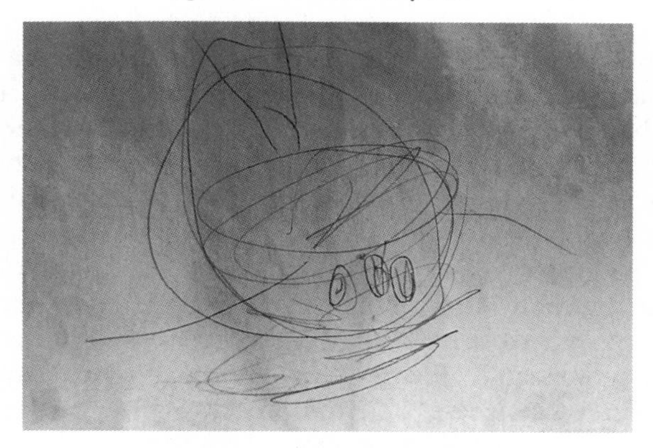

Figure 2.3 'Man not you'.

Figure 2.2 and Figure 2.3 are from the next session where B began to find tremendous reserves of courage in directly challenging her perceptions of men, using me as a sounding board to further clarify her understanding. Figure 2.2 frames an outpouring of aggressive and sexualised behaviour towards me: B shouted, 'This is you. Big eyes, big arms, big cock.' Somewhat taken aback by her directness, I offered her an interpretation that erred on the side of defensiveness by thinking aloud: 'You wonder sometimes B if I am like the person in the drawing?'

Along with the fears that were obviously bubbling away underneath, I believe that B was also attempting to tell me something about her discovery; that she experienced me as both penetrating and containing. It may have been more useful here to have said something like: 'Perhaps my bigness is not as frightening as it used to be?' Although her words carried a reparative intention, in the moments that she spoke them, I seemed only to be able to hear the fears that lurked behind them and which helped give them a negative edge. Hearing the positive intention over the negative tonality seemed a difficult lesson for me to learn at times, if only because of my occasionally surfacing resistance towards receiving projections of such a potentially ruinous nature.

In Figure 2.3, from the same session, B's concentration becomes more focused into her drawing. It becomes an image that both allows her to refine her distinctions between men that may harm her and men that will not, and to communicate her trust in me. Throughout the creation of this figure, B poked at me with her pencil. Here, B seemed to be putting into me ideas of both benign and malignant penis/pencil owners. She was communicating something of her need to be benignly penetrated by my words and interpretations, which served to help her to think more clearly and to discover whether or not I would have identified with the malignant penis/pencil owners and poked at her, as she sometimes feared I would. In other words, the pencil in this part of her play had the vital function of helping B explore ideas of benign and malignant penetration. In this sense, the pencil was her penis.[1]

In a later session, after B had further clarified that I was not going to hurt her by malignantly poking at her with my penis/pencil, she developed enough trust to look at a new image and her re-presenting trauma in more detail (Figure 2.4). This was an extremely painful session for me to witness, and the process of B creating this image contained all the terror of being sexually abused. Her plaintive cries of 'Help. Don't do it. Don't hurt me', whilst viciously stabbing her pencil into what she called her 'Minge' (the small heart-shaped area at the genital position), touched my own anger and frustration. Some of the feelings I experienced were due to the limitations of my professional role as an art therapist. Even though I had suspected some form of abuse and shared this with my colleagues, until this session there simply was not enough tangible material to request even an exploratory conference regarding her home situation. Having specialised in this field of work for five years, however, I have come to understand the absolute necessity of bearing 'uncertainty' and keeping an open mind in this most difficult work. It is all too easy to jump to premature and inaccurate conclusions in order to relieve oneself of the burden of not knowing. It is worth bearing in mind that to have one's fantasy reacted to as if reality, is undoubtedly as detrimental to a child's psychological development as is having her reality responded to as if it were fantasy. Exercising this knowledge in practice can be emotionally taxing for any therapist, especially when, in the developing transference, a child relentlessly imports powerful fears of intrusion and sexualises the relationship; and where the burden of responsibility for protecting a child, who is possibly being abused, lies with the therapist's sharing of subjective experience. It is essential,

Figure 2.4 'Don't hurt me'.

before embarking upon any one-to-one work with young children where sexual abuse is a suspected feature in their life, that whenever possible, one should seek to provide one's professional network with a supportive understanding of the nature of projection and transference.

Another issue, raised in this particular session, is that of the victim's incorporation of and differentiation from the abuser's violence towards the self. During her drawing, not only did B express the victim part of herself, she also expressed how the abuser had attacked her, and how this assault on her person had been taken in as part of her very own perspective on who she was and what she was about. There were times when she sought relief from being 'attacked' through the perverse enjoyment of attacking. If this particular fusion between the child's self and aspects of the abuser's personality are not expressed and differentiated, the child may continue to view vulnerability as an emotion worthy of persecution.

It would appear, from this case at least, that the very act of image-making well supports the processes of differentiation of self from internalised bad objects/people and their ways. The image is an object that comes from the self, yet gradually becomes separate. The child can take from it what she needs, and with the help of a thinking other, leave the rest behind. During difficult recollections, the art therapist can help the child find direction by re-framing loose anxieties or offering an interpretation; for example, 'You shouted for help, for it to stop, but you were still hurt and mixed up.' It is interesting to note that B's incidences of self-harm made a steady decrease from this session onwards.

In the same session, B made a drawing with which she attempted to piece together thoughts involving 'three men'. During the making of this image, B communicated what I had felt in the transference to be bodily intrusion by maintaining solemn eye contact, whilst pushing her fingers hard up her nose and down her throat.

In the child protection investigation that ensued, B was removed from her home, nursery and art therapy for nearly a week. This was the time needed to collate information from various sources, to assess risk in a rather complicated family system and establish her parent/s capacity for future protection. Sexual abuse was medically confirmed and the subsequent investigation resulted in the conviction of a babysitter.

Upon her return to nursery, B exhibited all the signs of fragmentation one could expect from a young child who, although having received protection, had also in the process met with an experience of having had every aspect of her life's continuity removed for a time. In addition to this, B also expressed confusion over the necessary physical examinations conducted upon her. During this time, she required much support in working through issues of separation, attachment and self-cohesion, and her sessions would typically include the excessive use of plastic sticky-tape, paper glue and cord in an attempt to give herself and her environment more continuity by sticking herself to it. B would usually choose a large piece of furniture or even a wall to attach herself to whilst crying for her mother. When she overcame much of her fear of losing her parents again, her preoccupation with sticking subsided.

In the following session, B drew and graphically explained how hair got into her eyes, nose and mouth, how 'dirty' this was and how 'bad' it all felt. She then picked up a small jug of water from the table and poured it over herself, saying: 'I want to change.' A female member of staff helped her to choose fresh clothing from the linen cupboard and to change what she was wearing. There followed a week when she peed herself and had to be changed. I shared my thoughts with nursery staff on how important it was to bear with B through this time, as her continued peeing and changing were likely to be linked with her need to let go of some experiences then inside her. In the weeks that followed, B worked with more thoughts, and there were many times when she needed to test my responses to her vulnerable side by symbolically offering it, for example by asking me to paint her, or poke her drawings with my pencil. During these times I needed to try and be clear about which of her invitations required a refusal to act, albeit symbolically, in a malignantly penetrative way, and which invitations were requests to experience my thoughts as benignly penetrative. It is important to note that when I had felt the former to be in operation, my refusal to act malignantly sometimes raised fervent anger in B, for not only had it provided her victim self with relief, but it also had challenged a part of her, which in those moments had sought to sense potency through perversion.

In our last planned session, B drew a small circle on a sheet of paper, and stabbed at it until its surface was ruptured. She then filled the space with glue and very carefully wiped the glue out with tissue paper, discarding what she called the 'Dirty' to the bin. After meticulously performing this ritual, B said, 'clean now', and made her way to play with the other children in the nursery.

One of the conclusions that can be drawn from this experience is that even though some aspects of the infant's and/or traumatised infant's art may not be at a stage of symbolic representation, it is nonetheless a highly communicable process, possessing a developmental function within the context of a facilitating human relationship. The infant-like re-presentation could be seen as a precursor of the more cultured symbolic representation, the main difference being that the former embodies and conveys raw emotional states sometimes associated with trauma, while the latter is a sign of certain 'raw states' becoming more bearable and thinkable for the child. The concept of a re-presentational mode of artistic communication can provide the child and therapist with an extra mental space, which is supportive towards the development of meaning in both its latent and, equally important, its manifest sense.

It would seem that within the therapeutic relationship, the created imagery not only has the potential to be viewed by the child as an external, carnate fragment of something that was once inside, but it also makes the route between the inner and outer worlds seem a more tangible pathway. Thus it assists communication between the two, and facilitates the potentially developmental processes of projection and introjection (Klein 1942). This could be an important factor, especially for children who are operating in a predominantly concrete way and whose main task seems to be the unmeshing of bad internalised objects with self, and, where the capacity to distinguish internal from external is limited.

Note

1 I am indebted to R.D. Hinshelwood for his insights into the complex nature of B's pencil play.

References

Bion, W.R. (1962) *Learning From Experience*. London: Heinemann (Maresfield Reprints 1988).

Fairbairn, W.R.D. (1971) *Psychoanalytic Studies of the Personality*. London: Routledge (1990).

Freud, S. (1910) Leornardo Da Vinci and a Memory of his Childhood. *Sigmund Freud 14. Art and Literature*. Penguin Freud Library. London: Penguin Books.

—— (1914) The Moses of Michaelangelo in Strachey, J. (ed.) (1974) *A Standard Edition of the Complete Psychological Works of Sigmund Freud*. London: The Hogarth Press.

Klein, M. (1942) *Envy and Gratitude and Other Works, Some Psychological Considerations*. London: Virago Press (1988).

Oxford English Dictionary (1989). Clarendon Press.

Pater, W. (1873) Studies in the History of the Renaissance. In Freud (1910).

Schaverien, J. (1987) The Scapegoat and the Talisman: Transference in Art Therapy. In Dalley T. et al. (eds) *Images of Art Therapy*. London and New York: Tavistock/Routledge.

Segal, H. (1957) Notes on Symbol Formation. *International Journal of Psycho-Analysis* p. 38.

Thomas, L. (1992) Some Re-Presentations and Transformations of an Abusive Experience, unpublished paper delivered to National Children's Home National Conference Working with Offenders and their Victims, Swansea.

Winnicott, D.W. (1958) *Collected Papers. Through Paediatrics to Psycho-Analysis*. London: Tavistock.

3 Arts Therapies with Children with Learning Disabilities

JANEK K. DUBOWSKI and JO JAMES

This chapter is concerned with arts therapies in the context of affective education generally and more specifically with reference to children with learning disabilities and special needs. This will be placed within the context of developmental psychology and Jung's theory and philosophy on the nature of the psyche. Affective education is defined by Sunderland and Clarkson (1994: 109) as: 'Centrally concerned with the understanding of emotions so that the participant develops conceptual frameworks for considering emotional experiences, one that equips her to deal more effectively and creatively with such emotional experience in everyday life.'

Affective education is relevant to every child's development, but particularly so for children with learning disabilities who often find it a real problem to articulate their feelings in words. There are further emotional consequences of this, as without a verbal vocabulary to express and communicate, fear and frustration (amongst many other emotions) may emerge.

Currently, the affective educationalist often appears in the role of therapist, and it would seem that the arts therapies and psychotherapies have amassed a wealth of knowledge and experience in working with feelings, which could be of value when shared with other professions. Affective education is concerned with emotional honesty and congruence to enable the healthy integration of the whole personality and well-being, and with emotional hygiene to ensure that the potential for unconscious and/or destructive 'acting out' is minimised. It is also concerned with emotional literacy to enrich the depths of meaningful contact with self, and empathic relationship to others and the world. As Sunderland and Clarkson suggest:

> With no education about the creative and effective handling of emotions people will tend to enter adult life emotionally illiterate, devoid of skills in how to cope with what they feel, and often chronically ignorant of creative ways in which to express and

> explore the emotions. Thus for many this deficit becomes a major obstacle in ever achieving inner peace and self-actualisation. (Sunderland and Clarkson 1994: 109)

Developments within the context of special needs and learning have been, and continue to be, in a process of change. Within the United Kingdom, the ideologies embodied in equal opportunities policies and normalisation philosophy have led to many children in this position entering mainstream education, whilst others with more severe special needs remain in special schools. Undoubtedly there are arguments for and against both such courses of action, and there are many complex issues requiring careful consideration in making these decisions. There is considerable anxiety, confusion and frustration amongst parents, teachers and children faced with the dilemmas inherent in, and surrounding, the process of learning in the context of learning disabilities. Resources to provide information and consultancy can be both erratic and limited. Consequently, there can be many casualties within the system whose needs are not being appropriately met.

An increasing number of children in this position are finding their way into relationships with creative therapists, within both the private and public sectors. These children are usually those who find it most difficult either to express or to contain their emotional reactions to situations, circumstances or people. However, historically, the therapy available to children with emotional or behavioural difficulties has been as a programme of repair rather than as a preventative measure. Therefore, it is perhaps necessary to consider whether the national curriculum alone is meeting the emergent emotional, social and psychological needs of our children.

Projects already recognising the need for affective education and offering creative therapies for children include the 'Place 2 Be', which now encompasses eight London schools; also the Islington schools project at the Institute for Arts in Therapy and Education and the Sesame Schools project in South London. These were initially established for so-called 'unhappy' children with special needs and/or emotional and behavioural difficulties; but other school children have also been offered the option of self-referring for help. An increasing number of children are self-referring, and there may be many factors that contribute to this increase, including under-resourced teachers, over-sized classes, and too narrowly focused educational aims and objectives.

The arts therapies, like any other form of therapy, are primarily a means to an end, not an end in themselves, and as such differ from case to case, patient to patient, and person to person. Different client groups might require specific forms of intervention and this is certainly the case when working with children, particularly children with special needs. One model, to be discussed later in this chapter (pp. 48–52), is often referred to as 'Developmental Arts Therapy' as it is informed by our understanding of human developmental processes. Other examples will be presented from an integrative arts therapy approach to affective education. One thing that all of these approaches have in common is that they offer children the opportunity to explore and express themselves on a symbolic level.

Symbols function on a variety of levels, and because of the particular qualities they possess, they have always been valued by therapists. For Jung, symbols came to be seen primarily as 'products of the unconscious' and as such provided a vehicle for exploring and working with the 'inner worlds' of his clients.

This symbolic process is observable in the spontaneous play of young children. As therapists we frequently observe this phenomenon. However we do far more than observe, we actively participate and enter into the client's imaginative landscape. Once we have entered this realm many new opportunities can emerge. It becomes possible to understand the client's experience, attend to psyche, and facilitate and effect change through the creative process, particularly as arts therapists are trained to be highly sensitive to the emotional content and psychological significance of metaphoric and symbolic expression.

During therapy, one of the methodological possibilities that can emerge to facilitate the therapist entering the child's world is that of role reversal, in which the child 'plays' the role of an adult and the therapist that of the child. During such role play, for example, the therapist might be told off for being naughty, fed pretend food, and found to be ill and put to bed. In order to comprehend what is being explored in such a session, it is important to be aware that the play contains both conscious and unconscious elements and manifest and latent meanings. On the surface, we can observe as 'manifest' in the play a child simply enacting a domestic scene. However, if we look more closely at the possible 'latent' levels of meaning, we might consider what this child is communicating about his or her own hungry, needy and vulnerable feelings that require attention. The creative medium becomes a channel for communication and expression on a symbolic level where literal discussion is not possible, and, as stated by Sunderland and Clarkson:

> The arts offer opportunities for the forming of feeling in ways that can enhance understanding. Expression through art media can

> often transform vague, confused or dimly aware feelings into communicable, clear and insightful statements ... young children's most fluent language for expressing feelings can be through the use of metaphor and symbol. Expression through art media can also be used to contain or free emotional energy as appropriate, to offer a safe channel for explosive feelings, to rehearse change and in doing so offer opportunities to try out new ways of being and doing. (Sunderland and Clarkson 1994: 112)

These assertions are illustrated in the following examples and case studies, which also highlight both application and process in creative therapy. The first concerns S, a girl of ten with a learning disability in a mainstream school who was sitting in classes that she could not understand and took away homework that she found completely impossible to do. Her self-image was rapidly deteriorating and so her parents sought therapeutic intervention from an arts therapist.

S complained of debilitating tummy-ache and had increasing experience of this in school. She brought this to the session complaining that she could not work or play that day. She felt immobilised. The therapist mentioned to her that school could sometimes be a very difficult place because so much was expected from her – it might feel awful inside sometimes. S's face lit up and after staring quite intently at the therapist for a few moments she announced that she wanted to play schools. The tummy-ache seemed to momentarily ease as S took on the role of teacher and put her therapist in the role of pupil. She started to point her finger and shout. She verbally attacked 'the pupil', stating her work was not good enough. The therapist, in role, felt her shame, guilt and humiliation – a process, sometimes described as countertransference, through which a client tells the therapist about how they are feeling through 'communication by impact' (Jung 1990a). S had used the dramatic medium as a vehicle for communication to tell of her experience. The inner debilitating self-criticism and anxiety about academic inferiority may or may not have been internalised directly from real-life experience. The therapist was now able to identify and to empathise. S cried and there was some apparent emotional relief. Her tummy felt much better. Perhaps her educational experience was unpalatable, indigestible and literally too much to stomach. In this example the art form had 'enabled' her, she had achieved something. She had found a 'learning ability' and with this comes self-esteem, self-worth and self-confidence.

T, a boy of twelve with quite mild learning disabilities and autistic tendencies, came from a special school class of sixteen children, some with very severe special needs. He was renowned for continued and increasing inappropriate behavioural outbursts and expressions of rage in the classroom. On such occasions it was difficult for the staff to attend to him and therefore he was excluded and isolated in a 'time-out' room alone until he could calm down. He was rapidly developing a very negative self-image and the difficulties were escalating. His mother was a single parent and was becoming increasingly worried and concerned about her son's welfare.

During the first session, in establishing a working alliance with him, the importance of privacy and freedom in therapy were stated. For example, all his feelings were welcomed. The therapist explained that sometimes it could be frustrating at school when there were so many children with different needs – perhaps he sometimes felt angry and fed up that there was no one to attend to his needs. It was explained that this was why his mother had decided to bring him to therapy. When asked what his own view of the situation was, T placed both hands firmly anchored on the arms of the chair and proceeded to bounce up and down on the spot smiling very gleefully. The therapist experienced in the countertransference an overwhelming sense of joy and relief. When asked if he would agree to enter into the arts therapy he leapt energetically up into the air exclaiming 'yes!' and jumped up and down on a pile of cushions with delight, demonstrating his emotional sensitivity.

Later in the therapeutic process, one that involved drama, music and movement, he dived beneath a blanket and made loud noises, telling the therapist he was a nuclear waste monster and going to contaminate them. He chased the therapist, who entered into the game pretending to be fearfully pursued, around the room. After a short while the therapist stopped and turned around, saying that perhaps it was sad for the nuclear waste monster that people were afraid of him and ran away, while actually they were rather fond of the monster and would have liked to have got to know him better. He stopped on the cushions and sat down, still disguised beneath the blanket. Perhaps he had a real fear of his feelings and therefore could not make authentic contact with them. The therapist began to touch his hands and feet through the blanket very playfully, suggesting that perhaps the nuclear waste monster was not as dangerous as he imagined. Here the therapist was communicating that they were not afraid of being in contact with the extent of his potentially destructive feelings. Before long he was laughing beneath the covers. He relaxed and lay down on the cushions. There was an experience of shared happiness in the room. The potentially explosive threat had been imagined and embodied, experienced and communicated.

It is undoubtedly true that people can and do respond very negatively to both children and adults with physical or learning disabilities, who may well feel rejected and unlovable – literally 'contaminated'. It might be easy for learning disabled children to internalise this rejection and feel that way towards themselves. Through metaphorical enactment it can be seen how the interpersonal relationship between therapist and client can catalyse intrapsychic change, as when T began to experience himself differently in the wake of the therapist's 'treatment' of the monster. Later during the same session, he lifted up the blanket and invited the therapist to go inside, explaining that they were both now inside the monster's tummy, as 'he has eaten us both up'. In this cavernous space together it felt both intimate and comfortable. He whispered that the little stains on the blanket were undigested food in the monster's belly. The therapist asked him if they were going to sit there and wait until digested, or what? He said no, and before long the monster was sicking them both up, regurgitating them out of his mouth and back into the room. T decided to fight the monster and was grappling with the blanket and a cushion. The therapist gave him a large bat and encouraged him to express his rage against the monster and accompanied the vivid physical struggle with a drum. Every time he hit the monster the drum attuned to the affect in the blow. He bashed the thing triumphantly to death and appeared satisfied and relieved when it was finally killed. 'That feels much better' he said, breathing out a deep sigh and smiling. He had been able to separate out from his overwhelming feelings rather than having remained a victim of them. He had grappled with them and this had taken a heroic ego strength of the kind required in the process of individuation. This was a simple example of how destructive energy can be creatively channelled and transformed in the process.

Some weeks later, again through dramatic improvisation, he found amongst the play sand a plastic toy snake with many little links like a chain. He decided to take this and pretend to strangle the therapist. This was not the first time he had begun to enact destructive feelings directly towards the therapist, who decided to formulate an intervention in response. Accordingly, the therapist told him she had noticed that he was often trying to attack or kill her in cruel ways and wondered why? She asked whether anyone had ever tried to do the same to him? He immediately stopped and looked very serious. The game was over and he was talking in an adult to adult manner in a very matter of fact way. He became able to speak about and describe in words the time when a gang of boys had chased him with a huge metal chain. The therapist, attuning to the feelings in this memory, suggested that it must have been really terrifying. 'Yes' he said, 'I was scared to death', and continued to recount the story in detail.

Here the art form had brought into consciousness a previously repressed memory, providing an opportunity to reintegrate potentially split-off parts of himself. If such memories remain unconscious they can have a detrimental effect or even be 'acted out' without any awareness. For example, rage attached to historical experiences can be transferred into the here and now. Perhaps the strength of T's fearful and angry outbursts in the classroom was a consequence of feelings from the past arriving in the present, without conscious realisation or knowledge of their original source. To change an effect we must return to the cause, and the arts therapies are concerned with the root of the problem rather than simply attempting to eliminate the symptom.

Later still in this therapeutic encounter and dialogue the therapist asked T if he would like to paint, and having spread out a large sheet of paper, T decided that he wanted to squeeze the paint bottles directly onto the paper rather than using a brush. His energy was high, and the therapist was concerned that he might get out of control in this process. The therapist's feelings could have reflected something of T's own anxiety about whether he could both express and contain feelings appropriately. Nevertheless he was trusted in his ability to do this and it was reiterated that here was an exciting challenge; the task being to keep the paint on the paper. The therapist was genuinely surprised that he did manage to do this, potentially echoing T's discovery. With tremendous energy and affect he squeezed out paint in vast quantities across the page, moving close to the boundary at the edge but never once overstepping it. When the therapist enquired about how this felt, the only word that he continued to repeat was 'great!' Having added many different colours, he sat down and with some satisfaction and pride, declared himself 'finished!' The therapist said that she was amazed at how well he had managed to both express and contain his feelings during this painting. He smiled and simply stated again 'great!'

Later in the session both moved across to the other side of the room and almost forgot about the painting in further discussion about his twelfth birthday the following week. The therapist empathised with him through a 'person to person' intervention about the difficulties of adolescence, being no longer a child and not yet an adult. Memories of difficult feelings characteristic of adolescence were shared by the therapist who gave voice to many conflicting, confusing and upsetting feelings that were inside 'when they were that age'. This seemed to really interest him enormously and he asked, 'What feelings? What did you feel?' The therapist replied with an image and said: 'Well I do remember that inside it felt very jumbled up – a bit like lots of tangled spaghetti – so many different feelings that it was hard to make sense of them and sometimes felt overwhelming.' He suddenly became very excited and said: 'Yes, yes. That's exactly how I feel', and rushing back across the room to his painting he continued, 'look, look at my picture – just like spaghetti'.

In this moment his isolation and loneliness were eased. Here the existential excitement and suffering inherent in this age and stage of development were shared with another person, were communicated, were understood, and therefore made somehow bearable through the creative process. The emotional world was consciously integrated and not unconsciously acted out as Jung (1990b: 210) suggests of the emotions: 'They cannot be integrated into consciousness while their contents remain unknown. The purpose of the dialectical process is to bring these contents into the light.' It is clear from this example how eloquently the arts serve this purpose.

After only eight sessions, the therapist received remarkably clear feedback from both the school and his mother that T's moods, attitude and behaviour had changed quite dramatically. There was a significant decrease in disruptive emotional outbursts reported and everyone agreed that he seemed much happier all round.

In beginning to work with children and young people who have more severe learning difficulties, Developmental Art Therapy can be particularly pertinent and potent. One of the aims of this developmental approach is to help the child to achieve his or her maximum potential. An understanding of *creativity* is as important to the art therapist as is the understanding of psychodynamics. The flexible problem-solving afforded by the creative process is one of the areas that lead to the claim that 'art' in itself can have healing properties. Visual problem-solving through the use of picture-making is something that children develop between the age of about eighteen months (when hand–eye co-ordination has developed to the extent that they can grasp an implement and direct it to a picture surface while attending to the activity) and four years, by which time most have developed the capacity to make recognisable 'pictures' endowed with symbolic meaning.

We take this very much for granted, but it is a complex developmental process to achieve the ability to work at a symbolic level with art. For persons who are developmentally delayed, as is the case with many individuals with learning disabilities, an understanding of these developmental processes is essential for the art therapist. These processes have been described by numerous authors who, while seeming to be in agreement that the following stages exist, have used different terms while seemingly describing the same events. For the purpose of this chapter, the terms used are those employed by Lowenfeld and Brittain (1974). The first stage is often referred to as *locomotor scribble* and describes the initial marks that are made when the child realises that the marker (crayon, pencil, etc.) leaves a permanent trace. At first children make the same movement of the hand repeatedly as if to reassure themselves that they are indeed responsible for it. The permanent trace provides a novel stimulation and this holds the child's attention for increasing lengths of time. This also improves hand–eye co-ordination leading to the next stage.

This is often referred to as *controlled scribble* and it is during this stage that the child begins to experiment with different movements of the hand leading to a variety of different types of mark. Rhoda Kellogg (1970) identified twenty different types of mark that she describes as the *basic scribbles*. These include single movements of the hand or arm leading to marks such as dots, single vertical, horizontal or diagonal lines, multiple movements leading to loops, zigzags or waved lines and roving lines that might incorporate enclosed spaces (as loops do). Eventually the child manages the more controlled and attended movements required to form a circle. It is also during this stage that the child can be said to be developing a 'vocabulary' of art as the repertoire of different marks increases.

The next stage sees the use of more than one type of mark on the same sheet of paper and Kellogg refers to drawings with two different marks as *combines* and drawings with several different marks as *aggregates*. Although these drawings are often exciting they do not yet resemble actual objects, as such; this stage and the previous stages are sometimes referred to as pre-representational as it is not considered that the child consciously intends to depict anything in these drawings. Some children do start to talk about what they are producing, however, and accordingly the next stage is sometimes referred as the stage of *named scribble*. When the child spontaneously points at what has been produced and names it, it marks a significant change in the child's thinking. In the previous *locomotor* stages the child is engaged in *kinaesthetic thinking* and once children name the scribble, they can be said to be engaged in *imaginative thinking* as they begin to connect their marks to the world around them, as in the process described by Lowenfeld and Brittain (1974). However, it is important not to impose one's own thoughts and ideas onto the child's work during this stage or to attempt to force this stage by naming the scribble *for* the child. This important stage marks a very complex change in the way that the child thinks and this cannot be hurried, as it is by allowing the child to play freely and experiment with drawing that will lead to this developmental shift.

Once the child has developed confidence with art materials and can play with them in a free and spontaneous way, the types of marks produced become more varied and/or complex. An interesting stage might occur when the child unintentionally produces a mark or combination of marks that actually look like something. When children recognise this they frequently become very excited. This stage was first described at the turn of the century by Luquet (1927) and named as the stage of *fortuitous scribble*, which he described as the child 'discovering the meaning of the picture in the act of making it'. According to Luquet this stage is followed by another called by him *failed realism* as the child cannot recall the movements of the hand used in making the recognisable forms. However, it is as if 'the penny has dropped', when the child realises that this scribbling can be used to produce something meaningful and now strives to attain this. During these earlier stages of hand–eye co-ordination, of attention and of concentration, a great deal of development takes place. It can be argued that aesthetic sensibilities, such as sensitivity to tone, line and colour can also develop during these stages. However, many people with learning disabilities do not progress to this stage and their use of art remains at the *kinaesthetic thinking* stages.

The next stage is sometimes referred to as the *schematic stage*, as at this time the child begins to combine various marks together repeatedly, usually in the same order. It is as if children are following a set of *rules* that they have developed for themselves. This provides a *structure* that allows for predicted outcomes and this can be very satisfying for the child or client. Even during the earlier locomotor stages the child has learned to follow some simple rules such as confining mark-making activity to the picture surface; is a rule imposed from outside, as few of us will tolerate mark-making on the furniture, walls or floor. However, there are other rules actually developed by the child, for example in maximising the novelty provided by making an exciting mark on a sheet of paper. This is a simple rule of 'do not mark where you have already marked' as if you repeat the same movement of the hand and bring the drawing implement to the same point on the picture surface the novelty very quickly stops. In order to follow this rule visual attention and a degree of concentration are necessary. These early experiences with rules and structures are further developed during the schematic stage when far more elaborate forms and structures are produced. As this stage usually occurs after the child has mastered the circle, many of these structures involve enclosures with markings on the outside, attached to the circumference or the inside of the circle. These schemata include what Kellogg refers to as *sunbursts*, circles with radiating lines that can be modified to include markings on the inside of the circle as well. In time these complex schemata are broken down until what we have is a circle with a few markings on the inside and a few radiating lines signifying limbs. This is the first recognisable form to be produced and is sometimes referred to as the *tadpole-man* and marks another major developmental step for the individual.

Until recently, the idea that prior to the stage of *fortuitous scribble* all mark-making is pre-representational had not been challenged. By pre-representation we mean that individuals have no conscious intention to link what they are producing to their experience of the world around them; their marks are not *intended* to look like something. Accordingly it has been regarded that those marks must be *meaningless,* hence the term *scribble* to describe them. John Matthews (1989) challenges these assumptions.

Matthews describes how children at these early stages are frequently seen to mutter or speak while they draw. He started to pay attention to these instances, and found that the children sometimes told themselves a little story while they drew, raising the question as to whether there is a link between what is being drawn and what is being spoken. Matthews demonstrated that on many occasions there was such a relationship, and he termed this form of drawing *action representation* in order to distinguish it from *figurative representation.* Matthews' work shows that prior to the development of figurative drawing (the development leading to the production of the *tadpole-man*), the child goes through a stage from *kinaesthetic* experience to one during which the line being produced takes on a representational function, such as for example a journey, and that this leads to the ability to 'later re-enact at a more conceptual level'. He concludes :

> Drawing is the only human behaviour in which a trace is made of actions performed. Children's drawings are literally recordings of their vitality and life. Denotional and aesthetic sensibilities to lines, shapes and colours probably comes about from identification and a total participation with the unfolding process of drawing, at a level when the multifarious expressive and representational potentialities are still in embryo. At this level the child is forming the 'double knowledge' (Furth 1969) that drawings are structures in themselves, yet simultaneously they refer to events and objects outside themselves. It is perhaps to this dimension of 'meaning' that therapists – and educators – should attend. (Matthews 1989: 140–1)

The capacity for symbolic functioning is necessary for human communication to take place, i.e. the ability to share the same meaning. This is a vital function for any therapist, and certainly art therapists attempt to achieve this through the client's pictures. If we consider that it is possible for individuals to develop pre-figurative symbol systems we must conclude that many people with learning disabilities may be *re-presenting* something of their experience of the world even though they may appear to be stuck at stages that have until now been considered as pre-representational. Such a model can have profound effects on the way that we view the work of such clients, and also provides us with opportunities to further develop intervention strategies for clients with communication difficulties, such as those suffering from autism and other pervasive developmental disorders.

Communication serves a vital function in the making and maintenance of social bonds and accordingly, difficulties with communication will almost inevitably cause major disability in individuals. Many persons with learning difficulties and language disorders will inevitably experience major difficulties associated with social development. This in turn can, and often does, interfere with ego development, insofar that the process of individuation is associated with the ability to differentiate *Self* from *Other* as a *social* process.

An example of this developmental approach in action is the therapeutic work carried out with G. A young person in his early teens, he had developed 'disruptive behaviour' to the extent that the special school that he attended was not able to contain him and accordingly he was excluded. G had severe learning disabilities, only limited use of language, and was autistic. The arts therapist started to see him shortly after his schooling had been interrupted. During early sessions he exhibited avoidance behaviour such as constant rocking while staring through a window, hiding behind his hands, covering his eyes when the therapist attempted to make eye contact, and echolalia. To begin with, it was quite difficult to get him to sit with some paper and crayons, but he would occasionally take one and move it around the paper. G did not look at what he was doing; instead he would rock his head rhythmically while looking upwards. During some sessions he would produce several markings made in this way, but was not able to concentrate on the activity or show an interest in looking at them with the therapist.

The therapist assumed a particular role in the sessions that followed, not only as a facilitator but also as a sensitive participant. Sitting or standing directly behind him and fully attending to his actions with the crayon, the therapist continued to describe with enthusiasm what was being witnessed. The therapist was participating in G's play and was showing him that his actions were important and interesting and worth attending to. The markings that he made were considered to be significant and of value. After some weeks this started to 'rub off' onto him. He too started to attend to what he was creating. His curiosity and focus began to increase as if he were internalising the therapist's active engagement with the markings on the page. Here the therapist was acting like an auxiliary ego, demonstrating and modelling contact and the relationship between the artwork and the creator, which in turn enabled G to allow this to affect the relationship between them. G had been shown how to value what he was learning, to value what he was making and to value himself. This led to increased confidence and the necessary developmental stages leading to representation. In time he became increasingly excited by the novelty of his own markings. G soon developed a wide repertoire of different shapes and forms until he mastered that very unique shape, the circle.

The circle holds the key to pictorial representation because of its property of containing. The circle has a boundary, an inside and an outside, and as a Gestalt is in itself a self-defining whole form. A circle can contain or it can differentiate, one can put other marks inside a circle or attach marks to its edges. The circle provides so many possibilities and once it is mastered it is returned to time and time again. The marks placed within the circle can become features: eyes, nose, mouth. The marks appended to the circumference become legs and arms, and later on ears, hair, hands and so on. Many researchers have suggested that this figure is the first 'representational' drawing that a child produces and as such we can see that the drawing begins to take on a 'symbolic' significance.

By the seventh or eighth session, G was producing rudimentary figures such as these and he was also demonstrating something else. These figures did not come from an external model, nor did they represent the therapist or some other external person. Instead when asked, 'where do the feet go?' G looked down at his own feet. When later asked about the belly button, he undid his shirt to take a look. He did what all children do at this stage in drawing development – he drew what he knew, not what he saw. And what he knew about the human form was what he knew of himself. This highlights the role of the arts in providing a tool for reflection and thereby self-exploration and discovery.

Another interesting thing accompanied this stage. Not only was G fully attending the drawing of these figures, the drawings became his. Earlier the therapist had difficulty in helping him to make any association with the marks he had made. Now after each session he wanted to take *his* drawings with him, and sometimes he made as many as thirty within half an hour. This was a significant milestone in ego development and therefore in his relationship to himself. Inevitably these changes occurring in the arts therapy sessions were transferable. His concentration and attention span increased, his disruptive behaviour diminished and he started attending school again.

Until an experience is expressed it cannot be shared; and expression therefore can be regarded as the first stage of communication. The case examples given in this chapter have involved children and young people who have managed to find appropriate means of expression and in some cases verbal language has been available as well as using art forms. If because of disability an individual does not develop language, the impossibility of adequately sharing his or her experiences with others can lead to frustration and more serious behavioural problems. At the most basic level, responses such as love and hostility require appropriate channels of expression, and if these are not available anxiety will inevitably result. In cases where persons have not developed linguistic skills, providing the opportunity to develop alternative 'non-linguistic' forms of communication can help to remove such blockages and the associated symptoms that these bring. The arts therapies – art therapy, dance therapy, dramatherapy and music therapy – can all provide such opportunities for expression; and art therapy, because it does not need to rely on the use of words at all, can be particularly useful with people who have severe learning difficulties.

This capacity to express oneself at the emotional level in non-verbal modes of communication has been the object of psychological study for some years, and perhaps the most important contribution in relation to this chapter is still the work of Susanne Langer (1942) and her distinction between discursive and nondiscursive symbols. The use of word meanings used in formal language, for example, is dependent on knowledge regarding the meaning(s) associated with each word (dictionary meanings, for example), and these are therefore regarded as discursive symbols that by their very nature are both verifiable and duplicable. Langer makes the distinction between this type of symbol and nondiscursive symbols that are far more personal, based on the individual's experience, and do not therefore follow a socially defined syntax and order. Beldoch summarises that:

> Nondiscursive symbols ... are communicated by their formal properties, and are the kind used by artists and in the process of intuition. This research suggests that abilities in the discursive and nondiscursive modes have some common variance, but that they are in many ways independent of each other. The implications for further research focus on understanding the correlates of ability in

> the nondiscursive mode, especially as they may relate to cognitive and conceptual function and malfunction. (Beldoch 1964: 42)

The use of recognisable pictures in art can be understood as a form of discursive symbolism insofar that we can all share in the recognition and understanding of what a particular form denotes. Nondiscursive images, as we have seen, do not rely on such conventions; the example of T's painting shows how he was able to express his 'very jumbled up feelings' in a nondiscursive use of painting.

Can the capacity for the use of nondiscursive symbols develop in the absence of a discursive system? In other words, can an individual who does not have the capacity to develop formal language, develop a personal nondiscursive symbolism? Moreover, if so, can we access this in a way that allows for a sharing of the meaning encoded in the symbol? By such a process can expression through the arts become an *alternative language* for these individuals? The examples in this chapter strongly suggest that such a process indeed takes place and that arts therapies, and indeed art itself, have a role in giving voice to children with learning disabilities who may otherwise feel lost, forgotten or abandoned by the system.

However, in the conventional classroom, the emphasis that teachers are required to maintain is often directed towards cognition and the thinking function. These can seem to be of primary concern and value, whilst the other capacities defined by Jung in his psychological typology as 'feeling, intuition, and sensation' may be ignored to the detriment of the developing child. In our logocentric culture in which the intellect is prioritised above the body and the imagination, there is a danger that emotional problems could increase, exacerbating if not incurring learning difficulties. It might thus be useful for teachers and therapists to collaborate in appraising and re-evaluating educational aims and objectives. We need to reinvent them in order to prepare young children, learning disabled or otherwise, for adult life.

It is sadly often the case that there are many right-brain children within a left-brain educational structure. In re-evaluating our priorities in preparing children for adult life, it is worth remembering that to have successful human relationships and develop a creative life requires emotional sensitivity, sensory awareness, and intuition. Yet within education these remain comparatively underdeveloped, given the emphasis on intellectual skills, and the needs for an affective education are unrecognised. In the meantime, therapy, and the arts therapies particularly, have a part to play in remedying this situation; not just as a means of dealing with the casualties of a deficient system, but as part of a collaboration with education in establishing the importance of, and contributing to, affective development. If our future culture is to be a humane one, then we need urgently to begin work at its roots, kindling and in some cases rekindling our humanity, integral to which is the process of attending to and attuning to the extraordinary emotional life of everyone within it. As stated by London Arts Therapists in Education:

> We see this as a critical moment in education, and not just in London. The new statutory guidelines on special needs, the code of practice, requires schools to meet all children's needs, even if this means buying in a service. And now that schools have their own budgets, they can choose to do just that. The National Curriculum Council has in fact recommended the use of Arts Therapies with children who have emotional and behavioural difficulties. We see it as part of our job to wake up schools to these facts. (Cohen 1996: 5–6)

References

Beldoch, M. (1964) Sensitivity to Expression of Emotional Meaning in Three Modes of Communication in Davitz, J.R. (ed.) *The Communication of Emotional Meaning*, New York: McGraw-Hill.

Cohen, J. (ed.) (1996) LATE (London Arts Therapists in Education) in *BADTh Newsletter* Spring 5–6.

Furth, H.G. (1969) *Piaget and Knowledge.* New Jersey: Prentice Hall.

Jung, C.G. (1990a) *Modern Man in Search of a Soul*, Reading: Ark Paperbacks.

—— (1990b) Anima and Animus, in *Two Essays on Analytical Psychology*. London: Routledge.

Kellogg, R. (1970) *Analysing Children's Art*. California: National Press Books.

Langer, S.K. (1942) *Philosophy in a New Key*. Cambridge: Harvard University Press.

Lowenfeld, V. and Brittain, L.W. (1974) *Creative and Mental Growth*. London: Macmillan.

Luquet, G.H. (1927) *Le Dessin Enfantin*. In Piaget, J. & Inhelder, B. (1969). *The Psychology of the Child*. London: Routledge & Kegan Paul.

Matthews, J. (1989) How Young Children Give Meaning to Drawing in Gilroy, A. and Dalley, T. (1989) *Pictures at an Exhibition: Selected Essays on Art and Art Therapy*. London: Routledge.

Sunderland, M. and Clarkson, P. (1994) The Role of Affective Education in Society in Trent, D.R. and Reed C.A. (eds) *Promotion of Mental Health, Vol. 4*. Aldershot: Avebury/Ashgate Publishing Co.

4 Shaping Vitality Affects, Enriching Communication: Art Therapy for Children with Autism

KATHY EVANS and MARIJKE RUTTEN-SARIS

In art therapy, the therapist and the patient, or client, are involved in a relationship in which they try to understand the art process and the product of the session, as an expression of what lies behind the issues that have brought the individual into therapy (Case and Dalley 1992). This process has evolved in recent years, as art therapy has moved towards a more psychodynamic object-relations approach. This has meant that for the art therapy to be effective, the dialogue that occurs within the art therapy sessions, whether it resides non-verbally within the artwork or whether it is eventually expressed in words, is one that has 'shared meaning' between art therapist and client.

However, working with autistic children, the specific communication difficulties they experience in both social interactions and in the use and understanding of spoken language, means that while an objects-relations approach informs practice, alternative methodologies are also necessary.[1]

The conventional Art Therapy Triangle provides a paradigm that identifies the three basic elements in the methodology of art therapy practice as the medium, the patient/client and the art therapist. Different methodologies will emphasise different dynamics within this triangle; for example, some emphasise different processes of interaction between art therapist and the client, while others emphasise the interaction between the client and the medium. In our work with autistic children, where there are severe difficulties in communication, the relationship between medium, art therapist, and client becomes more fluid and dynamic as the elements of the Art Therapy Triangle constantly shift in their relationship to each other. In working with children with communication disorders, the art therapist will interact and respond as an artist involved in the art-making process, besides objectifying this process by observation or analysis. This means that the art therapist is involved with the child and the creation of the art object or image with the empathy of an artist. This type of engagement, involving a particular sensitivity to and an identity with the very creative process of art- and image-making, can be described as 'staying with the image', a concept that has been of particular interest to McNiff who wrote of the image:

> The problem lies in our ability to open to it, attend to it, stay with and engage it imaginatively and with depth. We run away by trying to explain them according to abstract doctrines. The image is lost in the theory. The therapeutic cure and psychological depth involve the ability to stay with the image and perceive it with greater precision and sensibility. (1990: 28)

Thus, the art therapist, instead of keeping at a distance, becomes artistically involved in the interpretation of the emerging qualities and meanings of the artwork, and then is able to interact with it and with the child to give it further shape. This is a dynamic process, through which potential emerging meanings are tested and refined, and where interpretation is an 'understanding in action'. Within this process, the art therapist invests his or her own self, whereby readings of the artwork and the enactment of chosen meanings or values are not those of a dispassionate external survey, but are a commitment at risk, rooted, nonetheless, in a responsive and responsible sensibility.

The concept of the therapist being also an artist and staying with the image means that he or she also shapes the emerging process of the art therapy session, in the same way as they might creatively shape an artwork itself. The art therapy session develops with the therapist, sensitive to emerging nuances, forms and meanings, in a dynamic relationship with the autistic child.

In working with children with autism, the three elements of the Art Therapy Triangle are therefore interrelated in a holistic and dynamic way. However, there are further characteristics of this methodology that we call *Emerging Affective Art Therapy* that also redefine the nature of the art therapy triangle paradigm.

The art medium for example, becomes 'everything that is', insofar as within the environment in which the art therapy is taking place, the art materials, the child and the art therapist are intimately involved in emerging expressive and cognitive communication. As such, communication does not reside just within the finished product, but within the emerging and shifting dynamics of the whole art therapy process and its context.

The means that enable this dynamic to be perceived, recognised and developed by the therapist's interventions are located, on the one hand, within the notion of *Vitality Affects*, *Categorical Affects* and *Virtual Affects*, and on the other, within the cognitive developmental aspects of drawing and image-making.

Vitality Affects and Categorical Affects are part of the perceptual and communicative process that at the level of pre-verbal experience, provide the sensory foundations for such. Daniel Stern (1985) describes how our early infant sensory awareness of the world and our environment is formed by way of changes in sensation, or an *Activation Contour* in which existence, experience and activities occur simultaneously. Our experience is of a buzzing world, an undefined 'IT' that Stern describes as *Amodal Information*. However, in time, this abstract information becomes characterised by recurring rhythms, whereby certain sensory features become more defined and apparent. At this stage, however, the defined features are not necessarily sense exclusive, but are holistically perceived within cross-modal perception: a synaesthetic interaction among the senses, where qualitative dimensions can share sensory modes (for example, where changes in intensity involve things getting both louder and brighter, and where both sound and vision have an equal and interactive commonality in defining such a sensory change). The emergence of such perceived cross-modal features are called by Stern (1985) *Vitality Affects*, which then transpose easily into *Categorical Affects* as felt patterns of feelings, such as 'sadness', 'happiness', 'surprise' etc. The Categorical Affects and their emerging Gestalts, (recurring whole forms), become basic sensations and feelings to be further identified by signs and words, and thus as language. Stern (1985: 62–5) describes how innate abstract representation, by recognition of abstract patterns such as roundness and straightness, generally enables categorisation of abstract information. (For example, where changes in intensity involve sounds getting louder in one moment, and light getting brighter in another moment, both are experienced and decoded as a sensory change from small into wide and as a spreading sensation, giving rise to the Vitality Affect – spreading. The experience of many sudden spreading Vitality Affects, like a series of thunder, could find shape in a Vitality Contour and transpose into the Categorical Affect of anxiety.) However, before this refinement, Stern suggests that at the level of Vitality Affects , *Virtual Feelings* can be produced from others or other things; that in the presence of somebody else, we can 'feel' if someone is very 'excited' or 'sad' without verbal exchange. For example, in the presence of a patient who is reliving his or her spreading anxiety, the therapist can also experience Virtual Feelings of such anxious spreading, without verbal exchange. The patient's connection might be with thunder, the therapist's reality connection, on the other hand, might be with a passing train. They share the sensation, but not necessarily the same source or content. This process is not dependent upon conventional non-verbal communication, but upon the way our senses can cross-modally interact, and upon our sensitivity to cross sensory analogies and metaphors (as exemplified in poetic and expressive language, where we might describe someone with a 'loud' tie, or who is seen as 'green' with envy, or speaks with a 'bright' voice).

Within the context of art therapy, Vitality Affects radiate within the dynamics of the total and developing process, not just within the finished object. The sensitive and sensitised art therapist senses such Vitality Affects, encourages them as artist as well as therapist, and identifies the interrelated Categorical Affects, or feelings. In this way, pre-verbal communication can surmount the problems of the autistic child–therapist relationship.

For example, when D, a ten-year-old autistic boy, was painting, there were moments when the paint was being absorbed into the paper when he too became 'absorbed' in watching this happen. He was drawn into the total quality of the experience (Evans 1996), just as when we walk down a street we become absorbed into the crowd, or when reading, our thoughts become absorbed with what we are reading. As some of the children we work with like to tip the water onto the table and floor, we changed the paper to a more absorbent kind because it sucks in, or absorbs more water, and in doing so made the Activation Contour stronger for the child's experience; it heightened the context for Vitality Affects.

The importance of Vitality Affects to this total context is exemplified by further reference to the work of Evans with D. The art materials Evans took into the first session with D were coloured pencils and paper only. The setting was kept as minimal as possible to make the Activation Contour effective. The Vitality Affects conveyed by his first drawings (Figure 4.1) were perceived as *lightness, floating* and *flying*. The images, that were originally in delicate colours, are *sitting* on the edges of the paper (Figure 4.2) and the forms are *distancing*, which at the time seemed to be *inviting* Evans into a watching position. D, by having drawn a closed form, and constructing this perfectly, did not invite her in to share anything. A video of the sessions shows them, at this stage, sitting apart with D not looking at Evans, his arm placed like a barrier between them. The Categorical Affect conveyed by these Vitality Affects and his art-making were experienced as *sadness*. On an emotional and cognitive level, it could be sensed that he was asking ... 'Please don't come too close, I am

Figure 4.1 D's first drawing with floating Vitality Affect.

Figure 4.2 Another of D's early drawings, placed towards the edge.

coping with this but only if you are at a distance. I really want you to be there. All I want is that you be there for me so I can do this in my own way.'

In the second session, the images were *heavier* than before and more *consolidated*. There was more sparkle in the whole session and the art therapy was *settling*. On the video, they are seen together, with more glancing at each other. The movements between them were rhythmic, and towards the end of the session, D was almost sitting on Evan's lap. In the third session, paint was introduced to widen his experience. Keeping the crayons and paper as the prime materials, three bottles of liquid paint were placed on the other side of the table. During the session, Evans was able to indicate the bottles and D took them and squirted out the paint. He made five paintings/drawings during this session. The Vitality Affect was felt as *coming out*. The painting literally came out of the centre of several sheets of paper that were covered in a very sticky layer of thick paint. In another painting, he became involved once more with the brush and paint by flicking the paint around and moving the brush in a very distracted, dreamlike way. The Vitality Affects Evans experienced by D working this way were *absorbing*, and *scratching*. They made Evans feel that he was absorbed in what he was doing, although scratching because it was new and he was not used to it. The Categorical Affect she felt was one of *curiosity*.

A feature of later sessions was that D poured out a great deal of paint, and lifting it on the brush, intently concentrated on this. The paper became very soggy and thick with paint. The Vitality Affects felt during this session were that of *absorbing, hanging, pouring out*, and *lifting*. The Categorical Affect experienced by the totality of D's action and tenor, as well as by the artwork, was one of *uncertainty* and *attention*.

Later, D was encouraged to use a large sheet of paper, which made him stretch and reach out for the bottles on the table. At this stage, a feature of D's artwork was that he always started on the bottom edge of the paper when painting, and from there, moved outwards. Vitality Affects then felt were *reaching*, and *stretching*. The Categorical Affect felt was that of *control*; D directing his activity outwards from himself with more *confidence*. A suggested interpretation of this activity and painting is one in which D, feeling safe from this base, controls his uncertainty about things. This was also the case when he later paid more attention to taking the paint to the edges and filling the page, and when Evans felt the Vitality Affects to be *absorbing*, *soaking* and *immersing*, and felt their contingent Categorical Affect to be *satisfaction*; interpreted as 'I have done it' (something new).

The formal and qualitative way the art materials were used, rather than the content of the image itself, enabled Evans to sense Vitality Affects and Categorical Affects as the establishment of a pre-verbal, infant-based communication. This appeared especially pertinent as part of a developing process that qualitatively changed in accordance with the developing relationship between D and the therapist.

D, by experimenting with different materials and moving around in the art-making activity, entered the creative process in the same way as it occurs spontaneously in normally developing children at the pre-representational stage of drawing. In this way, he absorbed the ability to accommodate new elements into his art-making, and although his strategy for drawing or painting followed the same formula for construction, by widening his experience of art materials he was able to see his familiar representations, or schemata, in different forms. In spite of the rigidity of these schemata, which provided a safety net for his autism, this use of different materials and their different effects sensitised him to a potential capacity to symbolise.

Thus, this interpretative process within the methodology of Emerging Affective Art Therapy is further informed, not only by the active interaction of the therapist with the child, i.e. the process of 'staying with the image', but also by reference to the developmental processes involved in infant mark-making and its cognitive implications.

The awareness of the Vitality Affect radiating from the art therapy context provides a useful tool in the absence of the more familiar 'shared dialogue' that relies on spoken language or shared symbols or signs. Combining this with an understanding of the developmental processes of art-making makes it possible to assess the ongoing developmental aspect of the art therapy. Of particular relevance in our work are the very early stages of drawing development, the phase when the infant has first contact with art materials and art activity. Up until the 1960s, this stage of drawing development was more or less totally ignored in research on children's drawing. The abstract scribbles of infants were dismissed as meaningless, a phase to be experienced before the more 'important' goal of pictorial representational drawing was achieved.

Since then, however, several authors have pointed out that this stage is an important part of the development of art-making processes in the individual, if not the most important phase. It has emerged that complex planning strategies are required of children when drawing (Freeman 1980), and that even at an early stage, significant meaning is given by them to scribbles prior to the emergence of pictorial representation (Matthews 1984; Dubowski 1983). For example, a normally developing infant of one and a half years, when making the kind of marks shown in Figure 4.3 and Figure 4.4, named them 'butterfly', and 'mummy'. Such

Figure 4.3 'Butterfly'

Figure 4.4 'Mummy'

images, while pictorially non-representational, are nonetheless precursory to representation in that they are formed by actions that for the child have a reference to meaning (and hence are pre-representational rather than non-representational). For example, a boy of two made line after line with his crayon, accompanied by 'brr...brrr' sounds, resulting in a seemingly meaningless scribble. However, this was subsequently confirmed to pre-represent a car journey. Such pre-representations, in the form of scribbles and marks initially produced by 'playing around' with art materials, are by the actions of their making an important step in a developmental process towards an increasingly complex and representational ability to symbolise, and hence towards the development of language.[2] Significantly, Deloache and Marzolf (1992) speculate that experience of simple symbolic relations sensitises children to more difficult ones, and produced evidence to support the possibility that symbolic experience is cumulative.

For example, in working with D, Evans found that his drawings initially showed that he constructed images to a formula, i.e. a schema.[3] However, after fifteen sessions of art therapy, and widening his experience of art materials from crayons to liquid and hard block paints, it was possible for him to produce and see the schema in different forms. If this experience is cumulative, as Deloache and Marzolf suggest, then art therapy might increase D's ability to develop some, albeit basic, capacity to symbolise in spite of his autism. Shaping the art therapy process from his own 'Vitality Affects' into artwork, provided D with the possibility to engage in an interrelated, affective and cognitive process with the art materials and the art therapist.

The three main elements of Emerging Affective Art Therapy in working with autistic children involve the proactive participating creative role of the art therapist, the exploration of Vitality Affects and their Categorical Affects in reconstructing infant-based pre-verbal communication, and the developmental aspects of schemata in the artwork. In the following section we will exemplify this with a closer look at work that has developed from earlier research by Rutten-Saris (1991) on Emerging Body Language.

The method of art therapy dependent upon the recognition of Vitality Affects, their transposition as Categorical Affects, and their role in developmental processes, has been developed into a methodological paradigm by Rutten-Saris. This paradigm, termed as Emerging Body Language (Rutten-Saris 1990) is described in Table 4.1, and has evolved from analysing video-filmed sequences of infant care and art therapy over many years.[4] There are five categories, or phases, of *Interaction Structures* (displayed in the last four columns) in which Vitality Affects and Categorical Affects can be perceived and shaped within art therapy: *Attunement, Taking-Turns, Exchange, Play-Dialogue* and *Task*. Each describes the interactive nature of the child, therapist, artwork and art making activity. Each category of Interaction-Structure is characterised by a particular kind of *Intervention* on the part of the therapist that requires a response on a different and specific *Action Level*. Further, each involves or operates within an *Area*, or mode, of perceptual and cognitive engagement that ranges from the undifferentiated motor-sensory (formed or expressed by bodily movement) to the more conceptual (formed or expressed by symbols).

For example, within the first Interaction-Structure category, Attunement, the interaction between therapist and child is one in which the therapist picks up or becomes attuned to a Vitality Affect and its corresponding Categorical Affect. In working with this, the therapist's Intervention is one of *Being-moved-with*, or *Breathing in Rhythm* with the child, in which the required child's response is a corresponding action of *Being-moved-with*. The Area of perceptual-cognitive engagement is *Motor-Sensory*, characterised or expressed in body-centred *Movement*.

Although the five phases of Interaction-Structure are dynamic and interactive, in that they can overlap and change back and forth within a therapeutic session, or over several sessions, each can be seen as a particular developmental state of child self-awareness, corresponding to Stern's four 'Senses of Self' (Stern 1985). A fifth state, that of *Verbal Identity*, has been added by Rutten-Saris (1988).[5] For any one child, the

Table 4.1

© EMERGING BODY LANGUAGE (EBL) 1996

MARIJKE RUTTEN-SARIS

A diagram showing how the innate structures Rhythmic-whole, Abstract-presentation, A-modal and Cross-modal Perception, can develop into procedural structures of implicit knowledge that becomes visible in body-patterns.

Passive age /Active age	**Intervention**	**Area (Kephart)**	**Interaction-Structure**	**Action Level**
Stern: *I The Sense of an Emergent Self*		***Activation-Contour***	**Drawing level: POINT**	
Newborn P. 0-2 months/ A. 0-1 years	Being-moved-with Breathing in rhythm	Movement Motor-Sensory	Attunement	Being-moved-with
Stern: *II The Sense of a Core Self*		***Vitality-Affect-Contour***	**Drawing level: LINE**	
Baby P. 2-9 months/ A.1-2 years	Intense Moving-with	Body Perceptual-Motor	Taking-turns	to Move
Stern: *III The Sense of a Subjective Self*		***Categorical & Virtual Affect***	**Drawing level: CONTOUR**	
Crawler P. 9-13 months/ A. 2-3 years	Add **one** new element	'Image-like' Perceptual	Exchange	to Act
Stern: *IV The Sense of a Verbal Self*			**Drawing level: PLANE**	
Stepper P. 13-24 months/ A. 3-4 y.	Shape expectation Give names	Word Percept.-Conceptual	Play-dialogue	to Handle
V The Verbal Identity			**Drawing level: PICTURE**	
Toddler P. from 24 months/ A. From 4 y. on	Talk-About Give Task/Theme	Symbol Conceptual	Task	to Apply

category of Interaction-Structure reached has developmental implications and, overall, the Emerging Body Language paradigm may also be regarded as progressive. The interactions among child, therapist, art work and art making process, within each of the five categories of Interaction-Structure, develop from a more organic sensory and bodily orientated process to a more conceptually symbolic one. Once the Task phase is arrived at, the therapist's intervention is now verbal and task-setting and can relate to a more complex theme encompassing components from all phases. The Action Level of response now required is one of more complex bodily application, involving a symbolic or conceptual Area of engagement and is equivalent to Rutten-Saris' fifth state of sense of self, Verbal Identity. This developmental 'progression' is the key to a therapeutic process that is concerned with cognitive as well as expressive growth.

However, progression remains circular, and categories can cross over. Every part of the EBL matrix needs to be nurtured continuously, and in practice there is no main column or direction; every part or direction can be the most important, depending on the problems of the child or patient. In work with autism, for example, Being-moved-with and Movement can be regarded as a developmental achievement.

Within the Emerging Body Language paradigm, as the nomenclature implies, bodily and physical action and interventions are central to the shaping of the Vitality Affects that occur within the art therapy, their sensed Categorical Affects and the nature of the therapist's consequent interventions. This is particularly so at the earlier developmental stages, where movement and bodily action are both the means of interaction and intervention. In this process, the role of rhythm is important, particularly for Attunement. For effective Attunement to work at the first action level, a particular Vitality Affect has to correspond with a particular kind of intervention by the art therapist and with a certain Vitality Affect of the artwork and/or its making. The three actions must share some common currency that permits them to be transferred from one form into another. Rhythm is the strongest medium for this process. Every human breathes in a rhythm. Rhythm is an innate human feature that makes Attunement easier to perform given its universality, unifying qualities and efficiency (Leisman 1990). Rhythm can be delivered in, or abstracted from, sight, audition, smell, touch, taste and movement.

Before describing some examples of the Emerging Body Language paradigm, it should be stressed again that in order to perceive and feel the emerging Vitality Affect, the art session needs to be open, and with normal boundaries extended by the art therapist guiding the session with open goals. This is extremely difficult, as we are trained to educate children to keep the paint off the walls, to stop them from drinking dirty paint water, eating the crayons, ripping up the paper, sitting on paper, rolling in it, stamping on it, and so on. However, these are legitimate Vitality Affects within an art therapy context. They are all very spontaneous interactions of the child. They carry all kinds of Vitality Affects that could lead to many individual creative processes. *All* the child's actions are ways of working that shape art objects. Instead of offering them new ways to deal with what happens, we often only forbid that which they are already doing spontaneously. This is particularly so when they can talk or write, and when we expect them to understand our cues while at the same time being their spontaneous selves. For example, squeezing, soaking and dripping, biting and smashing can be shaped into ways of working with materials. Squeezed and twisted white drawing paper then becomes a sculpture, which can be carefully shown on a coloured board. The dripping-out of it can be caught on dry white pulp paper and turned into a colourful spot to become a painting. By biting, we can carve carrots, cucumbers and red and green apples into a colourful decoration. Smashing is necessary to work with clay to produce a smooth working material. The Vitality Affect to be worked on is often right before us, but becomes invisible because of our preoccupation and prejudices – we go out of tune, and out of rhythm.

The following examples of the EBL methodology at work were presented at the Leeds Uses and Applications of Art Therapy Conference, when they were illustrated by means of a video recording.[6] The first example took place when Rutten-Saris was providing EBL training for special teachers from schools for very multi-disabled children. The trainees suggested that to demonstrate the EBL methodology, she work with one of their pupils, B, an eight-year-old autistic boy. B had behavioural difficulties when confronted with adults, and heightened levels of social stimulation would lead to general arousal, which could result in the urge to masturbate.

The session with B that is described below began when B entered with his teacher to be confronted with the trainees. The session began at the Attunement phase. The Intervention by the therapist was through *Breathing in rhythm* and *Being-moved-with* the child. The Action Level response required by him was also *Being-moved-with*, and the perceptual cognitive Area was Motor-Sensory, as or expressed by movement.

As he entered with his teacher to be confronted with the group, B started to undress immediately and to lie down and excitedly spin around on the floor. Rutten-Saris, referring to the video, described her feelings and actions as follows (Evans and Rutten-Saris 1996):

> I feel scared and pleasantly tensed before this unexpected challenge. I get in a flash the interpretation that he will go to masturbate himself here right in front of us. That these adults around him stopped him without a cue for an alternative behaviour that would be socially appropriate to release his excitement. How do I now attune to B *and* to these trainees?

Rutten-Saris then responded to and interacted with B. She began to be *Being-moved-with* with B, and used the intervention *Breathing in rhythm*. Because she matched this quite well he was caught into a mutual shared rhythm. The first Interaction-Structure, Attunement, emerged. His spinning excitement functioned on a motor-sensory level and he became a little more relaxed. This provided the basis for an interaction and intervention that led to her responding to his bodily movements of spinning around as a Vitality Affect, sensed and transposed into a Categorical Affect of *shared excitement*. This in turn led to further interventions. She described this as follows:

> I am also immediately *Breathing-with* with B. Doing so brings me into the Vitality Affect of his rhythm. I reach a state of open active alertness, of sensing, smelling, trying and tasting. By doing so movements occur in time; an Activation Contour emerges. I start to enjoy the excitement of the whole situation. I catch his Vitality Affect of *spinning around*. My Categorical Affect and Virtual Feeling are of *excitement* about him and myself. I *Move-with* with it. At the same time I shape it rhythmically with movement, crayon and sound, first on the paper, and then on his belly.

Once the rhythm between B and Rutten-Saris had been established and the connection made within Attunement, the child started deliberately to *Move-with* the art therapist, who then used repetition to introduce a play of 'the same things'. A movement or action was repeated and then followed by a pause, and continued in a repeat–pause sequence, which was built up step by step to develop further connections and interactions between them. B thus became an active co-player, and in the terms of the ELB paradigm, the second phase Interaction-Structure, that of Taking-turns now emerged.

Every human breath has a rhythm, and within Attunement the intervention by the therapist was mediated through this. In the case with B, Rutten-Saris matched his rhythm and his Vitality Affect of spinning, by *Being-moved-with* him and providing an opportunity for a mutual Taking-turns behaviour. With B, it was the Intervention of *Intense moving-with* of body and crayon by Rutten-Saris to B's movements, and the spinning Vitality Affect that evoked B *to Move* himself more intensely in response. A structure was then created, in which both matched each other in half-minute sequences of repeated spinning movements and sounds followed by a pause. This led to the second phase Interaction-Structure, Taking-turns. The spinning play, from an energetic beginning, was gradually slowed down, with the pauses becoming longer and the intervening activity more structured and rhythmically patterned. The Action Level of response had thus moved from *Being-moved-with* to *to Move*.

B thus became involved in the triangle of himself, the artwork and the art therapist. As Rutten-Saris states: 'I caught and shaped his spinning so that he becomes able to answer the art therapist with a crayon on paper, with body movements and with sounds that are also socially acceptable.' Once the Taking-turns became more structured, she then introduced a sudden sharp scratching movement. By doing this and inserting one new element inside their now mutual play, the third category of Interaction-Structure, Exchange, could be explored. Accordingly, B actively accepted this new scratching element within their Taking-turns. The Action Level now required was *to Act*,[7] and so he too both repeated and inserted new scratching elements inside the now structured spinning. This was done with a mutual interaction and an increasing physical alertness, with B openly looking and smiling. He seemed pleased and happy.

Thus, to enable the child to explore the Interaction-Structure, Exchange, the intervention by the therapist is to add one tiny new element inside the Taking-turns that is related, as closely as possible, to the established Taking-turns format. The now established frame and structure of Taking-turns has to stay as a structural basis for the next, and fourth, Interaction-Structure, Play-dialogue. When within Exchange, B had expanded his playful actions with new elements inserted into the structure of the repeated Play-dialogue, he would have developed expectations about the start and end of these. By expecting a particular response, any small variation caused him surprise and amusement. This evoked Play-dialogue by way of mutual experiments. With B, this involved increasingly higher upward movements, accompanied by increasingly higher tone vocalisations, until at the climax of this, there was a sudden falling down of arms and shoulders, accompanied by a mutual looking into each other's eyes and loud laughter. Such actions can be accompanied by the therapist playing with verbal sounds and imagery, such as 't t t hee Rre..cooooooooomes a .....little....mOUSE..... into youuuuuuuuurrr........' Very often words will be used to shape a part of Play-dialogue. This might then be followed by some kinds of personal 'images' appearing. The child might use and understand simple concrete symbols like a sentence of two signs or words, such as 'want cat'. He or she could then *Talk-about* something, as in the next fifth stage of Interaction-Structure, Task. This could enable the emergence of conceptual thinking and the ability to respond with simple concrete tasks in response to requests such as: 'put your socks on' or 'draw me a cat, please' (responses that are nonetheless task orientated, even if in the case of the cat, a fully pictorial image is not achieved).

The Interaction-Structures, including Play-dialogue and Task, are further illustrated in the following example. G is a boy, who from birth screamed and cried, slept in short moments and was very precise about the food he liked. Although he had progressed through all the normal motor and language developmental stages and could talk, walk, read, write and calculate, he was unable to make judgements of height or distance, and had no feeling or notion of concepts like 'before', 'behind', 'under' and 'above'. As he lived on the seventh floor, his behaviour was extremely dangerous.

G acted as if he were a part of the space around him and responded to different spaces by extreme bodily movements and postures, and by strange vocalisations. For example, when in a corridor, he would become small and straight and vocalise in high tones. In an open field he would run, as if the devil were on his heels, and make low 'brumming' sounds. This behaviour suggested that he was experiencing the Vitality Affects of a space directly in his body. Further, any kind of movement made him jump up towards a door or a window. These patterns of behaviour were further complicated by the fact that locking him in had sometimes been necessary. At age six, he panicked when other children appeared, and at school it became impossible to contain him. He was constantly *Being-moved-with* with movements and screams by everything around him. It was as if the triptych mirror in the therapy room gave a correct picture of his repetitive expanding sense of self. He thus did hide-and-seek behind its doors and looked through the mirror in the same way as he looked through people. Rutten-Saris described her work with him as follows:

> I started art therapy with him when he was six years old without understanding one syllable of his Slovene speech. His behaviour consisted of loud sharp screaming, jumping, tiptoeing, flapping his arms, fixating through everyone, and opening and closing everything possible in the rhythm of his excitements. I attuned to him by running and screaming in his rhythm and shaped this into common children's play, such as hide-and-seek. Always I had my big sheets of paper and Stockmar wax blocks with me. I *Moved-with* with his spreading Vitality Affect of his running rhythm, by 'ticking' and 'lining' with the crayons on the paper and a blackboard.

Accordingly, when G ran around with wide-open arms and open mouth, Rutten-Saris produced within his running rhythm, soft curving lines with a start, an end, and a pause in the air. When he tiptoed, she ticked with the crayon point in a drip drip rhythm, the rhythm depending on the sharpness or roundness of his movements, and the length of the tick depending on the pressure in his body.

After two and a half years of intensive therapy and support, working through Interaction-Structures I and II, categories III, IV and V were entered into totally unexpectedly in the middle of one hour-long art therapy session. Rutten-Saris described this as follows:

> G invites me to come to play with him by 'lining' on the blackboard. I answer by *Moving-with* his 'lining.' For the first time then he non-verbally signs me to play hide-and-seek. An 'image' of the play has settled enough to recall it for his own purposes. G then hides me in the mirror. Suddenly, also for the first time he looks into the mirror into my eyes. Then he intensifies his gaze and looks intense into his own eyes, while naming by saying; 'look' for what he is doing. Because of G's initiatives I dare to go fully into the image area of phase III and use his 'lining' as a part of a 'creature'. After some trials from his side to get me back into 'ticking and lining' and Taking-Turns he suddenly shows Exchange on full image area; he draws a same kind of creature; a bird.

The session then moved into the fourth Interaction-Structure phase, Play-Dialogue, in which the Intervention by the therapist is to *Shape expectation*, *Give names*, the Action Level response is *to Handle* (both literally and metaphorically), and the perceptual-cognitive Area is now mediated more verbally. This Interactive-Structure is described with reference to G, as follows:

> G can name what he is doing and understands the direct needs of the scene (he cannot talk-about their symbolic meaning). There are 'images' and expectations. We enter the perceptual-conceptual area fully by communicating through the content of our images. 'The bird needs a door with a keyhole and a key in his cage.' During this image-communication he adds to his creature an inside lattice. I use this pattern to make an outside cage with a big door and a strong key. He answers by making small doors for himself and for me. I really feel sad that the bird cannot use the door to choose to stay in or outside the cage and I show this. G shows a clear emotional reaction by asking: 'Marijke?' He is worried, and wants to comfort me. He asks his mother with sounds and eyes for help.

She says, 'Marijke needs a big door'. He jumps up and draws a big door with a strong key for me. G completes the drawing by making also a big door with a strong key in his own cage.

The therapeutic session then moved into the final and fifth Interaction-Structure, Task. The Intervention by the therapist is *Talking-About, Give Task/Theme'*, the Action Level response required is *to Apply* and the perceptual–cognitive Area of engagement is now *Symbol* and *Conceptual.* Within this phase, once all the Interaction–Structure levels have been gone through, the child can make symbols, both within thinking and by talking. As Rutten-Saris pointed out, at this stage:

> There will be no need any more to check with an adult if he has done well or to look for what to choose or to do. There will be no need anymore to be afraid. From inside ideas emerge. Without hesitating the child will 'know' what to do; to do it, and to be creative.

She described what happened with G, as follows:

> Thus after I had finished the drawing, he deliberately went for a paper to cover the mirror and to shut the mirror doors. He gave himself a task, connected his problems without knowing about the interpretations made. After this session his behaviour lost most of the 'spreading' Vitality Affects. Only in stressful situations like a new school, or the war, did he suffer from it. *He was becoming a person using a space, instead of reacting to it.*

In this chapter we have set out to show how general patterns of behaviour connected to early infancy, the art therapist's own observation and analysis of herself, and a recognition that schemata in the artwork is important in art therapy for children with autism. The art therapist moving with the movements of bodies, hers and that of the child (Trevarthen 1977), and responding to the use of art materials to shape Vitality Affects, offers the opportunity of reaching children whose behaviour is less developed.

In this process, the first means of contact, intervention and communication, is through rhythm and movement. When people are together, they seem to be in a rhythmical dance. They are *Being-moved-with* (Condon and Sander 1974) with each other. It has been found that there is a connection between contact and shared rhythms with a direct influence on brain activities (Barthelemy, Martineau and Lelord 1992; Leisman 1990). Without some conscious control people cannot defend themselves from *Being-moved-with* unless they withdraw actively by becoming extremely self-connected or defensive like children with autism, as in the case of B and G. The group of people around B caused so much moving, noisy and smelly stimuli (Vitality Affects), that he could not defend himself against his emerging excitement and its inner-related 'spinning' tensions. The group's display was overwhelming for him. They took too much of his space. He *had to* masturbate (is *Being-moved-with*) to release his physical tensions (Rutten-Saris 1996). For G, the size of the space area related to the outwardly spreading tensions in his body. He was eaten by space. He *had to* (is *Being-moved-with*) run, jump and scream to release his physical tensions to be able to become a person in space. (The music therapist, Lehtonen (1995) also describes these kinds of tensions.)

B's reaction is just one problem that children with autism could have to face, which can make contact within the therapy difficult. Another difficulty is that they can have internal rhythms that are different and conflicting (Leisman 1990), for example, when the hearing of a sound over a long time precedes the 'Being-moved-with' reaction. It is also possible that the rhythm of the left side of the body does not accord with that of the right side, or that the upper part of the body is not in tune with the lower part. These disturbed 'rhythmical-wholes' make it difficult for the child to relate speech to thoughts. Such can make *Being-moved-with* difficult to establish within our therapy. 'Being emotionally moved' by your patient is often a very good start. No technique works here when you don't love the child as it is. To become able to enter a world of spontaneous expression of his or her feelings, the child needs to experience the feelings of his therapist as safe and alive, as pleasant, interesting and clear.

The EBL methodology and its processes do not have to involve going through the Interaction-Structure categories in a strict sequence from phase I to phase V. It has to start at the point where both the art therapist and the patient are. Their shared Vitality Affect will work through the persons they are at that time, the given situation and/or the artwork that they are then involved in. Vitality Affects work by themselves in what *is* at any given time. However, sufficient attuned work is required to enable the child to understand and to decide when and how to co-operate with whom. After this, it will then be appropriate to work with verbally given tasks and symbolic representations.

The therapeutic process described in this chapter begins with emerging Vitality Affects rooted in undifferentiated sensory and bodily modes of awareness, until through the recognition and shaping of their concomitant Categorical Affects an interactive relationship can develop. Through shared bodily rhythm and movement, creative play and art-making are facilitated by interventions by the therapist. Widening the experience of the children through art materials enables them to make their schemata and imagery more flexible, and makes it easier for them to cope with new information. As this becomes more differentiated and complex, their sensitivity towards symbolic experience is heightened, potentially leading to a more structured and conceptual self-awareness. In our work with autistic children, the Emerging Body Language paradigm is thus concerned with more than expression and emotional release, but with developing forms of human engagement, self-awareness and cognition.

Notes

1 For a more comprehensive view of autism see Aarons and Gittens 1992; Frith 1989; Trevarthen, Aitken, Papoudi and Robarts 1996; Williams 1996.
2 This is important with regard to children with autism, who have difficulty in developing a capacity to symbolise (Baron-Cohen 1992). However, with autism, the potential for this early pre-representational process to contribute to such a development is all too often 'closed down' by an intervention of abstract meaning, notably and usually by way of adult language. In Evans' on-going research the indications are that autistic children may 'cut-off' at the onset of drawing development, as soon as social interaction is required or demanded – for example, by the naming of a scribble by an adult (Evans 1997). The autistic child will then resort to and become 'stuck' in inflexible schematic drawing, which will not allow for the further development of pre-representational processes of symbolism.
3 Schemata are different from other images in that they are constructed to a formula: ordered in the way the lines follow one another, sequential in the way the forms are filled in, and with a conformity in the kinds of marks made (Freeman 1980; Kellogg 1970; Rutten-Saris 1991).
4 The EBL diagram shows a matrix that is derived from very different kinds of sources. Every column is derived from a different theoretical source or practical experience. The italics refer to the human innate cues that carry the development of some domains of the brain. These innate cues support their own development. One of their essential needs is interpersonal action, as described by the EBL diagram. In a healthy person these domains function as quite integrated. They are immediately intertwined because healthy infants sense their world as coherent. The word *Area* is used by Rutten-Saris because the EBL diagram can be applied in all directions for therapeutic purposes. The phases, or modes of engagement within Area are based on those of Kephart (1973): hence, for example, the term motor-sensory, rather than sensory-motor within the first category or phase of Interaction-Structure. The Action Levels, responses by the therapist and child or patient, are given in terms that are the nearest to their Dutch or German equivalents. *Being-moved-with* means being moved by inner and outer stimuli. *To Move* means that the child or patient repeats and increases a movement they like and stops one they don't. *To*

Act refers to the physical skills of action itself and involves a direct manipulation of something. *To Handle* means using appropriately the body or part of it, or a tool the way it is meant to be used towards a concrete future goal that is image-like, but not yet a fixed image. In *To Apply* there is a perception achieved of a future concrete goal that as such must involve symbols such as words, gestures, signs and pictures as action concepts leading to a prescribed goal.

5 Within each of the Senses of Self, there is a further categorisation of innate physical processes that is termed by Rutten-Saris (1988), as 'passive development' (P), corresponding to 'implicit knowledge' (Stern 1985) and 'procedural structures' (Dornes 1993) – and also acquired processes that can become available as skills and that are termed 'active development passages' (A), corresponding to Stern's 'explicit knowledge' and Dornes' 'explorative structures'.

6 The video film *Shaping Vitality Affects, Enriching Communication – Art therapy with Children with Autism* can be bought or hired from Department of Creative Therapy, Hogeschool Van Arnhem en Nijmegen, Holland.

7 Because this Interaction-structure, Exchange, is already somewhat 'image-like', the child needs a perceptual level of engagement to be able to Act, as indicated in Table 4.1 as the appropriate Area. This is equivalent to the requirements of Stern's phase of a 'Sense of Subjective Self'. This is also the Area in which Evans and D were working in the case study cited earlier. D, for example, accepted new elements, integrated them in his already existing schema and widened them.

References

Aarons, M. and Gittens, T. (1992) *The Handbook of Autism. A Guide for Parents and Professionals.* London and New York: Routledge.

Baron-Cohen, S. (1992) The Theory of Mind Hypothesis of Autism. History and Prospects of the Idea. *The Psychologist; The Bulletin of the British Psychological Society 5* 9–12.

Barthelemy, C., Martineau, G. and Lelord, G. (1992) *L'imitation du Mouvement* Congress, Arts et Medicine, la Recherche, AFRATAPEM, France: Universite de Tours.

Case, C. and Dalley, T. (1992) *The Handbook of Art Therapy.* London and New York: Routledge.

Condon, W.S. and Sander, L.W. (1974) Neonate Movement is Synchronised with Adult Speech: Interactional Participation and Language Acquisition. *Science 183.* 99–101.

Deloache, J.S and Marzolf, D.P. (1992) When a Picture is not worth a Thousand Words: Young Children's Understanding of Pictures and Models. *Cognitive Development* 7 269–72.

Dornes, M. (1993) *Der Kompetente Säugling. Die präverbale Entwicklung des Menschen.* Frankfurt/M: Fisher.

Dubowski, J.K. (1983) An Investigation of the Pre-representational Drawing Activity of Certain Severely Retarded Subjects Within an Institution, Using Ethological Techniques. PhD thesis, Hertfordshire College of Art and Design (now University of Hertfordshire).

Evans, K. (1996) Art Therapy and the Development of Communicative

Abilities in Children with Autism. Unpublished transfer Report to M.Phil, Faculty of Art and Design, University of Hertfordshire.

—— (1997) Art Therapy and the Development of Communicative Abilities in Children with Autism. Unpublished PhD thesis submission, Faculty of Art and Design, University of Hertfordshire.

Evans, K. and Rutten-Saris, M. (1996) Shaping Vitality Affects, Enriching Communication: Art Therapy for Children with Autism. Conference presentation, Uses and Applications of Art Therapy, Leeds Metropolitan University, May 1996.

Freeman, N. (1980) *Strategies of Representation in Young Children: Analysis of Spatial Skills and Drawing Processes*. London: Academic Press.

Frith, W. (1989) *Autism. Explaining the Enigma*. Oxford, and Cambridge Mass. USA: Blackwell.

Kellogg, R. (1970) *Analysing Children's Art*. Palo Alto California: Mayfield Publishing Company.

Kephart, N. (1973) *Hekkesluiters 1–2–3*. Rotterdam: Lemniscaat.

Lehtonen, K. (1995) Is Music an Archaic Form of Thinking? *British Journal of Music Therapy 9* (2) 20–6.

Leisman, D. (1990) Uit Ritme. *Tijdschrift voor Psychologie* November.

Matthews, J. (1994) Children's Drawings: Are Young Children Really Scribbling? *Early Child Development and Care 18* 1–39.

McNiff, S. (1990) Hillman's Aesthetic Psychology – From Psychoanalysis to Psychoaesthetics. *The Canadian Art Therapy Association Journal 5* (1) 27–38.

Rutten-Saris, M. (1988) *Ontmoeting in Beweging – Meeting in Movement: Emerging Body-Language and Music Therapy with a Mentally Handicapped Women*. VHS video film Project 3410, Bodylanguage and Social Care Niijmegen.

—— (1990) *Basisboek Lichaamstaal* (Basic Guide to Emerging Body Language), Assen: Van Gorkum. English translation, library of the Department of Arts and Design, Hatfield: University of Hertfordshire.

—— (1991) *Emergence of The Image in Art Therapy*. Hogeschool Nijmegen, Netherlands.

—— (1996) *Het lichaam 'vertaalt'. Het lichaam 'vertaald'* (The body translates. The body translated). Conference paper: Eeen andere kijk op het bewegen van ouderen. PAOG. April, Netherlands: Universiteit van Nijmegen/Hogeschool Nijmegen, Netherlands.

Stern, Daniel N. (1985) *The Interpersonal World of the Infant*. New York: Basic Books.

Trevarthen, C. (1977) Communication and Co-operation in Early Infancy in Bullowa, M. (1977) *Before Speech*. Cambridge: Cambridge University Press.

——Aitken, D., Papoudi, D. and Robarts, J. (1996) *Children with Autism. Diagnosis and Interventions to Meet Their Needs*. London: Jessica Kingsley.

Williams, D. (1996) *Autism. An Inside-Out Approach*. London, and Bristol USA: Jessica Kingsley.

5 Anorexia: the Struggle with Incarnation and the Negative Sublime

DAVID MACLAGAN

> One should call every sickness a soul-sickness.
>
> (Novalis 1956)[1]

Whilst the medical term 'anorexia' is just over a hundred years old, and descriptions of it as a recognisable syndrome can be found even earlier, there has been a good deal of recent debate as to whether or not it is actually a disease in any true medical sense. Like hysteria, anorexia has a vivid expressive, even theatrical aspect to it that is often difficult to contain within the conventional medical framework. Also like hysteria, anorexia has complex two-way links with contemporary cultural phenomena in art and fashion. Therapeutic strategies that identify the problem in individual terms, whether focusing on the body-image, seeking to restore a faulty *appestat* or exploring their situation within family dynamics can be effective, but there are aspects of the anorectic that escape such strictly clinical perspectives. Indeed it sometimes seems as if anorexia were designed to challenge the limits, and show up the limitations, of conventional therapeutic help.

A view of anorexia that confines itself to individual psychopathology without taking other dimensions, such as the anthropological, the cultural and the imaginal, into account will itself suffer a kind of self-imposed starvation. The lack of such perspectives does not only affect how we think about the 'therapy' of anorectics; it also thins out what the anorectic might have to say to us as a *figure*, in every sense of the word. As Angelyn Spignesi says, 'in the last century the hysteric carried Freud to psyche; the anorectic may be our twentieth-century carrier' (Spignesi 1983: 13). I take this to mean that the anorectic carries soul not only for herself, but for the rest of us, as I hope to show in this study.

I have chosen to single out one strand from a large body of pictures, about fifty in all, that were made by a young female patient, whom I shall call P, over a period of about eighteen months. My therapeutic work with her began in the context of a therapeutic community, continued in an out-patient setting, and wound up in private sessions, lasting about five years altogether. The pictures all derive from the early phase of this work, and some of them were made outside the art therapy group I ran.

I have deliberately chosen not to present a conventional case study, with its circumstantial details, because I believe that her imagery can be understood against a wider and deeper background. One of the advantages of reflective writing some time after the original therapeutic work is to enable it to be re-visioned, to follow up some of the links between the consulting-room and the cultural world in which it is actually located, and to allow the clinical to open out into the imaginal.

One reason why this might be the case is because the anorectic figure, in its struggle with the body, rehearses certain metaphysical conflicts that have a long-standing and peculiarly intense place in our culture. There are both religious and philosophical versions of these conflicts, but they have in common a deeply problematic and oppositional view of the relation between mind and body, spirit and flesh. In the context of this chapter, 'incarnation' is used to mean embodiment, not just because there happened to be a contrast in P's family between her Catholic mother and her father, or even because some of her art imagery specifically dealt with crucifixion and resurrection, but because her images expressed a struggle with incarnation that is both personal and archetypal. It is metaphysical insofar as it grappled with oppositions between spirit and matter, purity and muck, masculine and feminine, independence and dependence and life and death; and intensely human insofar as it involved a struggle to work a way out of these acute antagonisms towards some middle ground.

It is typical of the anorectic to live in a kind of liminal no-man's-land: on the edge between pre-pubertal and adult sexualities, between maternal and patriarchal authorities, between self-sufficiency and dependence. It is also edgy, in that it delights in taking things to extremes, in pushing them over the limits, sometimes with lethal results. These limits are often dramatically polarised, so that there is no middle ground, no place for compromise, only either/or.

It is here that the concept of 'soul' in its archetypal sense comes into play. For Hillman 'soul' is a quality of lived experience, rather than a metaphysical or spiritual entity; something that '... refers to that unknown component which makes meaning possible, turns events into experiences, is communicated in love, and has a religious concern' (Hillman 1975: x). Soul is also associated with deepening and refers to '... the imaginative possibility in our natures, the experiencing through reflective speculation, dream, image, and fantasy – that mode which recognises all realities as primarily symbolic or metaphorical' (Hillman 1975: x).

This archetypal or imaginal soul occupies a place somewhere in between the other-worldliness of spirit and the down-to-earthiness of the material world. It points beyond the literal or factual, yet it is also intimately dependent on an embodied life. Archetypal medicine has explored ways in which 'physical' illnesses, from cancer to rheumatism or asthma have a 'soul' dimension to them: for example, the work of Zeigler (1979), Kugelmann (1985), or Sardello (1983). These ideas are not just about physical illness having a psychological aspect or psychological disturbances having physical accompaniments; they are about an imaginal or 'soul' dimension to illness.

A psychotherapy that adopts an imaginal perspective, that sees symptoms, life events and case history as images rather than matters of fact, will find it easier to pick up and tune into signs of soul. Art therapy, both because it deals with actual pictorial images and because its inclination is towards the imaginal, is especially suited to do this.: It also has a special responsibility towards the aesthetic.

It is important to be clear about what 'aesthetic' does and does not mean in this context. Aesthetic qualities are not purely formal, nor are they evaluated in terms of some idea of 'beauty'; on the contrary, they can convey the whole spectrum, from exaggeration or ugliness to the kitsch or the indifferent. Despite Freud and Jung's treatment of them as distracting or self-indulgent, they are intimately charged with psychological sense. The significance of art therapy pictures, and indeed of any picture, is not to be found only in their figurative or symbolic aspects (the 'what'), but also in the way they have been made, their style, their material facture (the 'how'). In other words, there are intimate and complex links between their aesthetic and psychological aspects (Maclagan 1995).

This has a particular relevance for anorexia, in some cases at least, because aesthetics plays a crucial part both in the anorectic lifestyle and in its pictorial imagery. In a sense, anorexia embodies an aesthetic of extremity: this is evident not only in the obvious paring down of the anorectic body and the horror of 'fat', but in many other details of the anorectic lifestyle, including dress, furnishings, and even intellectual 'furniture', in which the environment is subjected to stringent control. Heywood, for example, has recently written about the parallel tracks – academic and athletic – on which she forced herself to attain physical and mental perfection (Heywood 1996: 2–9).[2]

This control is not exercised just for the sake of control, though it is a fierce territorial assertion; it can also be an attempt to substitute a reliable and material support system for the insecurity of any close personal relationship. Such an environment is like a solipsistic or narcissistic version of Winnicott's notion of environmental provision, where a person tries to provide for themselves that which it is too dangerous to depend upon others for. Such a rigorously defined territory can function like a psychic immune system, attacking and expelling foreign bodies, in the name of a perfectionist and idealised aesthetic. And it is at the edge of this extremity, at those impossible points where attempts are made to realise the ideal, that we encounter what I shall call the 'negative sublime'.

By 'sublime' in this context I am referring to a cluster of qualities that are at once aesthetic and psychological. We are familiar with sublime imagery from art of the Romantic period, and Kant made it an important part of his transcendental system. These qualities connect with feelings that are fascinating, awe-full and even terrifying: they seem to point beyond the realm of the ordinarily human. The sublime is also at the edge of what can be represented by a work of art: its imagery points towards what is beyond the reach of symbolisation. To that extent it displays its own limitations, hence its attraction for some postmodernist thinkers. Although the anorectic sublime does sometimes cast a shadow of illusion or lack of substance, it is certainly not to be identified with postmodernist disillusion, for in it the shortcomings of the real are pitched against an immense and powerfully attractive ideal, that has an archetypal or cosmic tinge to it. In psychological terms, the negative sublime is a demanding image of impossible perfection whose influence on ordinary life is corrosive and potentially lethal.

To give a sense of this anorectic sublime, at once fascinating and terrible, I would like to cite one of the images made by P (Plate 1). It shows a great white bird ominously poised over a landscape of desiccated black trees (a typical feature of several of her images), below which are the tiny box-like dwellings of men. The feeling of colossal indifference, the discrepancy between the huge, all-embracing, white hovering other-worldly creature and the pathetic ticky-tacky boxes of the human world, is overwhelming. This image seems to me to convey vividly a kind of inhuman, or superhuman, yearning for something so utterly 'beyond' that it is actually hostile to the ordinary world, and it conveys a powerful feeling of the uncanny aura of the negative sublime.

Anorexia presents us with paradoxical juxtapositions between indulgence and starvation, a cupboard of hoarded goodies and a ruthless self-denial; between a claustrophobic closet and an intolerably public world, between ecstasy and excess and a correspondingly fierce need to ration and control. What has to be refused is too tempting, too rich and too dangerous. In one of her poems, P wrote:

> I cannot bear the silence,
> I cannot say the words,
> I cannot feel the shapes.
> My words are silent,
> like daggers,
> like pain, like spikes inside of me,
> like stuffed-up, shut-up words,
> to stuff-up, shut-up you.
> Silent eyes,
> in the cold and open space,
> in the cold and open sky,
> too big to hold onto,
> too big to see,
> too painful to feel.
> I cannot look at you,
> I cannot meet your eyes,
> nor meet your heart.
> It's very lonely in here,
> with too many flowers and trees.
> too much love,
> too much passionate enterprise in me ...

The nourishment that the anorectic refuses is not simply food, to be accounted for in material terms (fat, calories, or vitamins) or even in terms of its material textures and flavours (chewy or soft, harsh or bland); physical starvation should, rather, be seen as a concretised metaphor for other forms of self-deprivation and control (Spignesi 1983: 17–19; Levens 1995: 109–10). But it is in the nature of the condition to refuse to de-literalise this perspective: it is as if the anorectic is bent on presenting a problem that is quite literally incorporated, and this insistence reinforces the obsessional and secretive rituals that make her so difficult to treat.

These difficulties are not just due to psychopathology; or rather, their pathologised aspects do not belong solely to the anorectic individual, nor even to the psychodynamics of her family: they belong to a much more universal mode of imagery. Pathologising, in Hillman's sense of an active and dramatic image-discourse of affliction, deformity and morbidity, that is peculiar to soul, resists being roped into a merely clinical corral:

> The wound and the eye are one and the same. From the psyche's viewpoint, pathology and insight are not opposites – as if we hurt because we have no insight and when we gain insight we shall no longer hurt. No. Pathologising is itself a way of seeing: the eye of the complex gives the peculiar twist called 'psychological insight'. (Hillman 1975: 107)

Anorexia presents us with an image that transcends a merely functional or behavioural perspective: its enactments, and the images it presents jump out of the boundaries of therapy and ask to be understood against a more general, cultural backdrop. In other words, the contradictions that are acted out in anorexia, and the pictorial images that sometimes result, cannot just be made sense of in local terms: the resonance is a collective one.

It is obvious that there are powerful empathetic links between the anorectic and the prevailing images of fashion and the media. At first there seems to be a dramatic contrast between anorectic starvation and the wealth of consumer imagery. But this 'consumption' has a double edge: on the one hand, there are images of luxury, of narcissistic indulgence; and at the same time these images, particularly those of female models, exhibit a sophisticated leanness, an immaculate reduction to the bare minimum, or a waif-like deprivation. Hence anorectic starvation can be seen both as a reaction against the display of conspicuous consumption and as a secret collusion with the spectre of poverty and emptiness that haunts it.

Furthermore, the idiom of what might be called the cutting edge of recent advertising imagery is one of transgressing boundaries, or operating in a liminal area: between male and female, normal and perverse, choice and addiction (to name only a few). The image of anorexia is incorporated in the culture that it seems to parody: like other modern icons of 'pathology' (sexual abuse, drug addiction, etc.) it migrates far beyond the clinical domain, in very much the way that, almost a century ago, hysteria circulated between the clinic and the arts (dance, theatre and literature).[2] The explicit connection between the clinical dramatisations of hysteria in Charcot's public Salpétriere sessions and Loie Fuller's dance imagery, has only recently been explained (McCarren 1995).

Another related theme revolves around the motif of the degradation of the image of woman. This is something more fundamental than the distortion or abuse to be found in the realm of consumer imagery. Here it is the association of woman with the natural, the material (a word that some etymologies connect with *mater*, mother) that is problematic. On the one hand we are confronted by idealised images of cosmetic woman as immaculate and streamlined; on the other there is the periodic and messy reality of menstruation or the irregularities of physical 'imperfections'. There are also all the stereotypical associations of 'femininity' not only with attractiveness, but with passivity, self-abnegation, etc. These are some of the ways in which anorexia, or at least the anorectic figure, can be understood in non-clinical terms: as an icon of transgression or defiance in cultural, or even in political terms. Some of the multiple cross-overs between religious fasting, anorectic starvation and political hunger-strike have been explored by Ellmann (1993: 13–15).

One of P's later pictures gives a vivid, and somewhat mocking, picture of this (Plate 2). A pure white phallus stands up against a 'ground', a tangle of hair, flesh and blood, in which her protruding hand shows her to be embedded. But the monumental phallus is too clean-cut to be true, so that its proud masculine authority feels rather hollow or unreal, and there is a sense of life and creative fertility in the embodied earth with which she is identified. One could also suppose that this picture presents a caricature of the antithesis between a certain stereotype of the sublime and the mundane material world against which it is pitched, but without which it is lifeless.

Many of these paradoxical conflicts also connect, in a distorted or 'unconscious' way, to deep-seated and collective fantasies about 'goods'. This complex bundle ties together several factors that are rehearsed in anorexia, in ways that are *critical*, in every sense of the word: the relation between materials and qualities, the hiatus between glamorous appearance and inner emptiness, and the ambivalent role of aesthetic decision. These are factors that feed into what I have already called the *negative* sublime. Both situationist and postmodernist critiques have pointed out the 'spectacular' or facsimile aspect of consumer culture, and indicated many of the ways in which appearance is substituted in it for reality, and in some cases this has led to a cynical or nihilistic flattening of all gradations of value (e.g. Baudrillard 1985: 126–34).

There are several, extremely different, ways in which the figure of anorexia relates to these contradictions between surface and depth, image and reality. At one level, anorectic starvation can be a horrifying literal riposte to the consumer imagery of extravagant plenty. In this embodiment, fantasy and fact, the symbolic and the actual, are collapsed together. At another, more distanced, level, it can also be a dumb-show, a silent scream of protest. At this point, its meaning cannot be translated unequivocally, and anorexia plays upon the edge of what can be comprehended: the baffling or paradoxical nature of this riddle, and the very real obstacles to therapy that it poses, have an ironic inflection to them. Irony, and its sardonic relatives, are characteristic features of anorexia: it blurs the line between failure and refusal, sometimes acting out the old grammar of 'I shall drown and no one will save me' to the lethal limit.

The hiatus between the material and the ideal, which is often given a typically adolescent twist in anorexia, is also crucial to what I have called the problem of incarnation. Embodiment involves various forms of compromise, the toleration of imperfection and the holding-together of literal and symbolic in a metaphoric equivocation rather than their collapse into concrete symbolism. Nowhere is this more evident than in the anorectic relation with food. The danger is, for both the anorectic and her therapists, that the threats to which starvation is a response – the calories or the lost pounds – will be taken at their face value rather than treated as *images*. To do this we need a metaphoric perspective, both on starvation and on the nourishment it rejects.

The actual process of self-starvation has a long history, which includes fasting for religious purposes. Such practices were means as well as ends: not only did they demonstrate, often to public view, spiritual ambition and subjugation of the flesh, but along the way they often produced desirable side-effects. As the business of eating was reduced to a minimum or virtually eliminated, the mind was less distracted by bodily appetites; concentration and clarity of perception increased, and the body itself eliminated impurities. This reduction or purification of bodily processes and substance was thus both an end in itself and the means to an end – the elevation of mind or spirit over matter or flesh. Here again, we encounter the potentially lethal connection between sublimation, the sublime and an impossible Beyond.

Even religious starvation involves a mixture of sacred and mundane motives: an asceticism designed to discipline the flesh may become a strange form of tribute to it, the boundary between punishment and worship is blurred. This is much more obvious in modern dieting or fasting undertaken for health reasons: here there is a secular cult of the body, but with metaphysical undertones. Indeed it is possible to argue that there is a continuous spectrum from the stigmata of self-inflicted suffering, through more obvious narcissistic shaping of the body (including cosmetic surgery), to both female body-building and the apparently opposite extreme of anorexia.[3] All of these practices entail rigorous editing of the body, in the name of a transcendence of material imperfection that is as much aesthetic as it is spiritual.

P made many images of cleaned-out, empty containers (bowls, acorn-cups, shapes almost like the inside of hollowed, emptied breasts): these were holders of immaculate emptiness (Figure 5.1).

Their aesthetic quality is sometimes witness to this: under their apparently innocent, decorative surface, there is the sense of a purified perfectionism; it is as if any internal obstacle has been strictly eliminated, and what is left is an absence whose ominous character is camouflaged by cheerful colours.

Figure 5.1 Acorn cups.

Related to these images is a remarkable picture that is a kind of celebration of this potentially lethal pruning. Against a gold background stands a carefully smoothed black 'trunk', whose relentless purity is underscored by gold lines (Plate 3). Only the red sections at the top of this truncated trunk suggest, in a rather shocking way, the presence, under the immaculate surface, of something else. The whole picture looks and feels like a strange sort of talisman, the object of a cult. And this is exactly what it is: an icon of the worship of disembodied being, but a figure whose semi-disembodied abstraction cannot avoid conjuring up what has been sacrificed in the process of its creation.

Anorectic starvation is both a sacrificial offering and a figure of deprivation. Needs for sensuous pleasure, emotional warmth or material comfort that are not met can be denied, and this renunciation can then seem like a sacrifice, an offering at some chillier, less human altar. The divisions between physical and psychological, which are reflected even in the variety of therapeutic approaches to anorexia, also appear within anorexia, as we have seen, as a struggle against the implications of being embodied; a rebellion against the compromises and ambiguities of incarnation, with the muddle in the middle.

This anorectic ambition to escape the body and its organic untidiness produces its own mirage of a discarnate mode of being, a clarified, frictionless and weightless Beyond that has a glamorous allure and an attraction, which can prove ultimately fatal. This lethal and ethereal Beyond is the unearthly, dehumanised version of a much more real sense of the gap between an almost ecstatic experience of exuberance and a sensibility that can only tolerate this torrent of feeling in tiny doses. Once more, the sublime appears as something ethereal and transcendent, that is at odds with the ordinary world.

Two of P's pictures illustrate this. The first shows a somewhat ornamental blue bird thrusting up into the sky (Figure 5.2). The bird's body is in fact a hollow bowl, of the sort I described earlier. The clouds

Figure 5.2 Ornamental blue bird.

through which the bird makes its passage have a plump cotton-wool look to them, as if they are not quite real. She said that the bird could just fly on and on for ever, there was nothing to weigh it down or hold it back. I think we can see this picture as an image of sublimity that has a false feeling to it: the ease with which the empty bird cuts through space, its streamlined aesthetics, and the abstract feel to the space itself, offer us clues. An aerial cruising, a sublime ambition, is depicted, but it is something that has quite lost touch with any ordinary world, that is suspended in a Never-Never Land.

One of P's poems is clearly about this picture, and it begins:

> The great blue, electric bird shimmers in the sky;
> Forcing a path through the thick, white blankness,
> Piercing cloud, through to cloud, through to cloud,
> Flying endlessly on with its dead-weight of emptiness,
> searching eternity.

In the next stanza this relentless, ethereal ambition is reversed, literally turned inside-out, in a ghastly return of the repressed:

> Until, one day, one hour, one minute,
> it can bear to stop,
> and turn on itself,
> and plunge its own sabre beak deeply into its own cavernous body,
> to spill and feed on its own entrails.

That this contradiction need not be so sterile is shown in a second, later, picture (Plate 4). This shows the excruciating discrepancy between an almost unbearably lush, colour-saturated world (at left) and the bleak, dark wall of deprivation, behind which P depicts herself, peering through a tiny chink at the radiant beauty on the other side. Here something that could be called sublime, in the sense that it exceeds comprehension, is represented *alongside* its obverse, the shadow world of self-deprivation; and we are allowed to see the fragile, minute point of contact between them. The latter is crucial, from a therapeutic point of view: it shows that the patient herself is able to tolerate 'seeing round' her own starvation.

The difficulty in accepting incarnation, the capacity to live with or inhabit the body and its untidy processes, rather than subjecting it to persecutory control or mortifying metaphysical ambition, seem to be problems peculiar to anorexia. But the anorectic, like the hysteric before her, may be functioning somewhat as a 'condenser' of dilemmas or crises that haunt our culture. What we find in the context of therapy cannot always be understood simply within the confines of therapeutic theory – defective body-image, faulty *appestat*, familial disempowerment or whatever: it belongs to another dimension, one to which notions of unconscious causality or personal history are marginal. This dimension could be called metaphysical, 'archetypal' or 'imaginal' according to choice: what matters in this context is how it invites a re-visioning of the sublime.

The eerie aspect of the sublime is well known from archetypal psychology. Jung described the feel of 'visionary' art as:

> Sublime, pregnant with meaning, yet chilling the blood with its strangeness, it arises from timeless depths, glamorous, daemonic, and grotesque, it bursts asunder our human standards of value and aesthetic form, a terrifying tangle of eternal chaos ... (Jung 1984: 90)

I think the problem with Jung's description in relation to anorexia is that there *is* an aesthetic, but that it is an aesthetic that is paradoxical:

it revolves around a detailed scrutiny that is extremely particular (in every sense), and at the same time it has an anaesthetic effect, in that the sensuous and emotional apprehension of the world that is implied in the etymology of 'aesthetic' is strictly limited, or kept literally 'out of touch'.

Hillman has taken 'aesthetic' back to an almost phenomenological level, rather than to its erstwhile association with rarefied and precious styles of experience:

> The link between heart and the organs of sense is not simple mechanical sensationalism; it is aesthetic. That is, the activity of perception or sensation in Greek is aesthesis which means at root 'taking in' and 'breathing in' – a 'gasp', that primary aesthetic response. (Hillman 1981: 31)

The absence or numbing of such fundamental aesthetic response leads, in his view, to collective as well as individual pathology. It is as if in anorexia there is a characteristic deformation of this response.

I have tried to show how some aspects of the anorectic aesthetic can be made sense of in terms of a peculiar version of the sublime. Whilst this sublime is out of reach, it is at the same time paradoxically literalised and aggravated. This ungraspable aspect of anorectic deprivation is conveyed vividly by a fragment of Kafka's:

> Those hardest to please are certain ascetics who practice hunger-strike in all areas of life and thereby seek at the same time to attain the following:
>
> 1. a voice will say: Enough, you have fasted enough, henceforth you may eat like others and it will not be counted as food for you.
>
> 2. the same voice will simultaneously say: Thus far you have fasted under obligation, from now on you will fast with joy, it will be sweeter than food (yet at the same time you will actually eat).
>
> 3. the same voice will simultaneously say: You have defeated the world, I relieve you from it, from eating and from fasting (at the same time you will fast as well as eating). (Kafka 1961: 189)[4]

The twists and turns of self-persuasion and the manipulation of bodily needs that this text presents are like a surreptitious, almost private version of the more extreme forms of self-denial that are a tribute to the negative sublime, but they involve many of the same paradoxes. What Kafka's fragment also shows is the anaesthetic effect of starvation, that ends up insulating the fasting person from any contact with reality and hooks them into a sublimatory spiral of metaphysical aspiration in which they become increasingly ungrounded.

How might art therapy work with this? As we have seen, the first mistake in therapy is to be drawn into interventions pitched at the same level of concreteness at which the anorectic operates. We need to recognise the imaginal or soul dimension behind even the most alarming reality. Because art therapy by definition works with images, both literally and metaphorically, it is in a position to avoid being misled in these ways. But even so, the authority of the psychodynamic tradition is hard for us to dispel, particularly as it is often presented as the reliable backbone of art therapy training, while the psycho-aesthetic or imaginal approaches that were a crucial part of art therapy's original heritage tend to be at best taken for granted, and at worst effectively disqualified.

We need to try to respond to the distress and discordance of the anorectic in terms that are not simply psychological or functional. The perspective of true therapy is one of attending, meditating, following and deepening rather than a medical one of battling with, or eliminating, sickness. It involves something more like what Thomas Moore calls 'care for the soul' (Moore 1992). In any case, as I have tried to show, the anorectic figure beckons us beyond the confines of the consulting-room. Much of the work of understanding the role of the sublime in anorexia is retrospective. It has taken me years of living with and 'handling' the images my patient made, to begin to appreciate their richness, and to realise – because these are 'pictures' in several senses – how much they give onto a world of soul. We have much to learn from the anorectic figure, not just as therapists, but as human beings.

Notes

1 Author's translation.
2 For more general cultural aspects of hysteria, see Micale 1995, Chapter 3.2.
3 For a look at some of the covert collusion between anorexia and female body-building, see Heywood (1995: 35–40).
4 Author's translation.

References

Baudrillard, J. (1985) The Ecstasy of Communication in Foster, H. (ed.) *Post-Modern Culture*. London: Pluto.
Ellmann, M. (1993) *The Hunger Artists: Starving, Writing and Imprisonment*. London: Virago.

Heywood, L. (1996) *Dedication to Hunger: the Anorectic Aesthetic in Modern Culture*. London: University of California Press.
Hillman, J. (1975) *Re-Visioning Psychology*. New York: Harper.
—— (1981) *The Thought of the Heart Eranos – Lecture 2*. Dallas: Spring Publications.
Jung, C.G. (1984) *The Spirit in Man, Art and Literature*. London: Ark.
Kaffka, F. (1961) *Parables and Paradoxes*. New York: Schocken Books.
Kugelmann, R. (1985) Breath, Speech, Asthma and Psychological Understanding. Paper presented to Human Science IV International Research, conference, University of Alberta.
Levens, M (1995) *Eating Disorders and Magical Control of the Body*. London: Routledge.
Maclagan, D. (1995) Fantasy and the Aesthetic: Have They Become the Uninvited Guests at Art Therapy's Feast? in *The Arts in Psychotherapy 22* (3) 217–21.
McCarren, F. (1995) The Symptomatic Act, *Critical Enquiry* (21).
Micale, M. (1995) *Approaching Hysteria: Disease and its Interpretation*. Princeton: Princeton University Press.
Moore, T. (1992) *Care of the Soul*. London: Piatkus.
Novalis (1956) *Ausgewahlte Schrifte*. Hamburg: Fischer Bucherei.
Sardello, R. (1983) The Suffering Body of the City: Cancer, Heart Attack and Herpes, *An Annual of Archetypal Psychology and Jungian Thought* 145–64 Dallas: Spring Publications.
Spignesi, A. (1983) *Starving Women*. Dallas: Spring Publications.
Zeigler, A. (1979) Rheumatics and Stoics, *An Annual of Archetypal Psychology and Jungian Thought* 19–28 Dallas: Spring Publications.

6 Exploring the Landscape Within – Art Therapy in a Forensic Unit

KARL TAMMINEN

Art therapy does not only prescribe a way of working, but describes a way of seeing. As an art therapist I have been given insight into other individuals' views of their lives, world and ultimately themselves, in a way technology is unable to provide; no scan, no surgical technique, no camera can allow us access to another's view or memory of their life. However, through the use of images created by individuals, we are given access to a unique view of the world that tells of formative factors, developmental influences and physiological changes that control, effect, create, enhance and change our view and ultimately our understanding of the world.

Every individual has a unique view of the world; no two can be the same, because no two people have been in the same places, seen and done the same things. Just as no two views are the same, each interpretation and understanding of the world is right for each person, at that time and in that place. It is often impossible for us to understand how we can see the world so differently from others; but then we are not looking through their eyes, nor have we lived the way they have lived. Art therapy, however, gives us the opportunity to capture glimpses of the world through the eyes of another.

Because the human brain stores memories in different ways, we have to be aware of the present understanding of the process by which we retrieve these memories, an understanding that is evolving, but is nonetheless helpful when applied to verbal and non-verbal processes. The left-hand side of the brain, which stores logical, sequential and language-oriented experience has difficulty interpreting, processing and expressing the information stored by the right-hand side, which specialises in perceptual and spatial experiences. Much is lost when we try to express or work with feelings in a totally verbal way; what there are not words for, we cannot express. When we try to retrieve right hemisphere memory through left hemisphere techniques that are primarily verbal, we are often left speechless. Memories and the feelings they carry are rarely orderly, logical and precise; and so art therapy, through the utilisation of creativity, which is a more primal means of expression and communication, attempts to create a forum in which these aspects can be explored. The utilisation of creativity allows the use and development of a language that not only can access, but can explore and work with the non-verbal memories and feelings that are the 'landscape' of an individual's sense of self.

An understanding that sees people as individuals with their own unique perspective and experiences but also in a social and historical context, further informs my practice as an art therapist within a medium secure forensic unit with people who have the label 'offender'. This understanding allows me to see the individuals I work with as people, rather than as a collection of behaviours and criminal activities, which would by necessity preclude them from society. The role of an art therapist on a forensic unit is complex and many-sided; there for the individual, to establish a strong therapeutic relationship, to use skills to support and facilitate, and to respect clients and the time spent together. Yet beyond this relationship with the individual, there has to be an awareness that there are other expectations of, and layers to, a role that is constantly brought to mind by the presence of alarm systems, airlocks, security and restriction. To work with people who have the label 'offender', no matter how distasteful or horrific the crime may appear to oneself, or to the society that has removed them to a place of safety and security, there is a need to return to this core understanding in working with an individual's sense of self. There is also the need to acknowledge their identity as more than a name or genealogy, but as a jigsaw of events, interpretations and understandings from both within their life span and before. Similarly, it is important to remain constantly aware of the need for both therapist and client to retain a sense of self that is defined and informed by parents' and peers' expectations and their own experiences, and limited by our own biological and neurophysiological ability to interpret the world in a tangible sense. The world is what our senses and interpretations tell us it is. We are whom we perceive ourselves to be, but our perception has been shaped and clouded by the very process of living.

A medium secure forensic unit is the end of the line, the only other destinations being the regional secure and special hospitals. It, therefore, becomes a make or break situation where an individual makes a decision to change, on whatever level and within their ability to do so, or to remain the way they are. *Informed consent* is a crucial aspect, which reflects the very nature of art therapy itself. Informed consent is not about the words we use or the replies we receive; it is about establishing common terms of reference, our understanding of meaning and our ability to understand others and express ourselves in the form of a chosen communication.

For forensic clients, the decision is often not solely about whether they wish to participate in therapy, but whether or not they wish to be part of society. For forensic clients to return to society, a society they often view as hostile and uncaring, either the society has to change or they have to believe that their perception of society is not the only one possible. Therefore, informed consent is the ability to accept that there are alternative ways of understanding ourselves and the society of which we are part, and to further acknowledge that such perceptions and roles are ultimately flexible and can be transformed and changed. If an individual can give informed consent, or can move towards such a position and in turn make the decision that he or she wants change, then therapeutic co-working can begin. Ultimately, with both therapist and client constantly aware that change, perceived or measurable, is observed by agents of society (including the therapist as such), therapy can help in the client's decision as to whether to live within society or outside of it, and whether to exist in a reality that is collectively defined, or personally constructed.

In the area of forensics, the therapist is constantly aware that there is a difference between the view of an individual *with* a problem and an individual who *is* a problem. Labelling is not helpful: it is a description that describes the exterior or presentation but does not give a clear indication of the workings or the cause. Art therapy in a forensic unit cannot be about surface appearance; it has to dig deeper. It has to chart uncharted territory, and it has to become part of this internal landscape, if only to make it visible in a seeing world. But that is just the beginning. Separating the offence from the person and not working with a person who is offensive involves painstaking navigation. Too often, people become their problem and are not differentiated by the outside world as individuals *with* a problem.

Out of necessity in working with forensic clients, the offence is recognised as the end of the journey, and I backtrack with the individual to the cause. We discover together how an understanding of the individual and their world and all the component parts that include moral code, values and attitudes such as racism and sexism, is formed, changed, lost and reconstructed, and how experience forges opinions and understandings which ultimately shape and mould our sense of identity. During the journey we learn things we are never consciously aware of or able to express in words. Ultimately, all experience can guide the review and any change that can result from such self-exploration. Something of this process is illustrated by the case that follows.

Arrested for arson J, the subject of our example, denied it and spoke about his confusion over why he was detained. While on remand, awaiting his court date, he became depressed and tried to commit suicide. He was diverted from custody and sent to a secure ward for psychiatric assessment, and his story unfolded over three months, leaving both the individual and the therapist reeling at the lifetime of memories and feelings, hitherto stored but denied, verbally inaccessible, and presumed lost, that exploded from the images that he created.

It is not possible at the beginning even to hazard a guess at the direction and course therapy will take; the complexities of a lifetime of experience make this at best, impossible and at worst, foolhardy. This 36-year-old man was arrested for numerous arson attacks on upper-floor flats and maisonettes. The case was complicated by the discovery of pornographic and erotic drawings that were compilations from memory, imagination and adult magazines, and intended to depict the women who were living in this targeted accommodation. Society, and the legal judiciary system it employs, wanted to know if this man was responsible for his actions or if he required the help that mental health systems can offer. Diverted from the courts for an assessment, the client, distressed, confused and afraid, wanted clarification and help.

The images produced by J, within the art therapy sessions, are visibly accessible and they have a technical skill which is not always available to all clients who participate in art therapy. In *spite* of this there is a strength in the images that goes beyond skill, and it is this strength that has a quality like sand slipping through my fingers, which I am trying to capture.

The first image that J created in session was a self-portrait standing next to an unnamed woman, a self-portrait of how the individual then saw himself and wanted to be seen. The second image he created was of a childhood memory, one amongst many, providing a feeling and understanding that cannot be fully verbally expressed. Through the adult's memory of a scene, depicted through the child's eyes, we were able to witness his first day at school, how he had wandered into the girls' playground, unaware of doing something that was not allowed. The older girls were hitting and teasing him, and present in the background was the adult who would take him to the head teacher for punishment, at the very moment J thought he had been rescued and the torment stopped. This memory and these moments had become part of the way J saw himself and his place in the world.

There were no words for this and no single image can encapsulate the details, but his image captured some of the feelings. We could have used words to try and explain and express, but the image had already fulfilled that need in ways words could not have done. Any words used at this stage were about establishing a common frame of reference between us, and so I asked, 'Is this what you wanted to make me feel and know? Did you experience what I experienced, or do I remain unconnected to those around me? Am I really as isolated and distinct as I believe myself to be?'

The image-making provided the means for J to externalise childhood memories, for example depicting seemingly ordinary scenes such as Christmas. Nonetheless, it began to raise key questions. Why had J chosen such an image; why, for example, was one figure of a more substantial quality than the others that were apparently transparent? Again words were used to begin to gain common terms of reference: 'Could you tell me a little about this image?', to which he attempted to respond and put this memory into words: 'This is my memory of a good time, there are only five of us because when we reach the age of fourteen we have to sit outside the room and listen ... Christmas is for children ... I liked Christmas, I do not know why you can see through every one but me.'

The delicate threads and connections and subtle interplay of seemingly unconnected factors, like tumbling dominoes pursuing many directions simultaneously, were triggered and highlighted by the production of these childhood memories and by the questions they raised. Art therapy provided the means and the opportunity for a release of painful memories, and the feelings, thoughts and behaviour they enveloped. The initial images had hinted at an apparent surface ordinariness, but this illusion was swept aside by the client to reveal the complexity beneath the surface through this new form of communication that made something without form or shape suddenly visible. Memory, feeling and understanding, which had waited thirty-six years for a forum and opportunity that would make their visibility permissible, now provided the necessary environment for self-reflection, learning, self-appraisal and change at a fundamental level. Transformation and change are not the same: change can be adaptation, and in art therapy within a forensic unit, as with any other location, there is potential for the fundamental change that can transform the way individuals see themselves and the way in which they are seen by others.

Abuse hidden, disguised and denied, will make itself known when the opportunity arises. In Figure 6.1 the excrement on the abuser, the child, even the shadows reflect not only the material he fell on, but also his perception of the act in which he was left feeling tarnished and unclean. What we feel now, as we look at this image, we should trust. A common frame of reference has been established, and of the many people who have been stunned by this image, all reported the same response in terms such as 'it is like ...' and 'it makes me feel as if ...'.

Figure 6.1 'I've never told anyone this before.'

Figure 6.2 'How people have hurt me.'

One week after having disclosed his abuse, while trying to communicate through an image measuring six feet by four feet (Figure 6.2), we began to piece together the jigsaw of how J saw himself. This was not a child; this was an adult held down, helpless. This was not fire; this was semen, burning his eyes and burning misplaced shame into his very sense of who he was. His world was chaotic, and family, friends, and enemies had all become unrecognisable within an ocean of angry expressions and frightening, unreal eyes. 'I am an adult but I am also still burgled, still mugged, ultimately I am still eleven years old, part of me has not moved on.' J's words keyed us into reference points along with the narrative images. For example, with reference to another image depicting a child trying to attract the attention of a female adult who is busy at the kitchen sink: 'Afterwards, I tried to tell my mother, but she didn't even turn, all she would say was "you dirty disgusting boy ... go and get washed and changed". Looking at the pictures now, do you think she meant me? Meant the abuse? The dog dirt on my jumper?' In this forum there are no absolute truths, only a truth for this moment. In this place and for this person, clients can begin to question how they may have been seen and how they see themselves. His self-questioning was crucially important: 'all those years so secretive, furtive, drawing women and fantasising, wanting to do to a woman what had been done to me ... why? Why should I want to hurt a woman when it was a man that hurt me?'

There followed a sequence of six images, done over three months, in which the abuser is depicted as a monster: the unknown abuser who had destroyed his marriages, his friendships, had alienated him from family and ultimately and fundamentally left him impotent, both sexually and in terms of self-determination. The need to destroy the monster was evocatively depicted in the series (Figure 6.3).

This sequence of images embodied the belief 'if I destroy the monster, everything will be all right', a belief, shared but not questioned or further explored, as this was a time for just watching and listening. J's words sought to clarify a developing narrative: 'In my imagination I am a hero, I save people I do not hurt them ... It fights back but I am strong, I am winning, I can destroy the monster.' While I was on annual leave J

Figure 6.3 'I can destroy the monster.'

produced this image (Plate 5) and his commentary was accusing, 'I was alone, no one supported me, the beast has grown stronger, I am going to be destroyed.' Suddenly we had been given a new frame of reference, as pieces interlocked for J. What had been felt, seen and known by him was given tangible form and verbal expression: 'perhaps that is why I am screaming'. I sought explanation and understanding, my frame of reference required elaboration and restructuring, and J replied: 'it must be true, it feels true, they are both me ... I am good and bad, I set the fires, not the man who hurt me.'

After this shift in understanding, J was reeling with the effects and knowledge of too many layers of his sense of self. His words rang emptily and were blurred with confusion, but in Figure 6.4 there were the beginnings of a realisation: 'I'll have to control the part of me that wants to do bad things.' In those few words, and particularly by the use of the word 'control', there was not only a recognition of the effect of actions; but rather than transformation or fundamental change, a simplistic theory of control is applied and ironically is generally expected. After a week of thought, this idea was then referenced in a further more figurative image verbalised as follows: 'I've been thinking, I can't kill this monster, it's me, I'd be killing myself, I can't chain this up for ever, if only I could turn the monster back into the child.' Insight in its truest sense is fluid and part of an enlightening process where each understanding has an effect on the next.[1] There is not absolute insight amidst the complexity of human sentience. Here was an indication that feelings and understand-

Figure 6.4 'Controlling the bad thing.'

Figure 6.5 'Fire is love.'

ing were changing, shifting and reforming, although words alone reflected confusion and perhaps even confused further. The musing that accompanied a later image was more focused: 'If I could just join the child, let him know that it was not his fault.' There was now a common frame of reference between J and myself, but we struggled to verbalise this understanding. We could express our belief that if the client did not blame himself, blame the child, and feel the guilt, he might be able to feel, remember and accept the memory of abuse as his own childhood memory without hating and denying the child, as if it were some separate entity. Then he could have begun to feel complete rather than fragmented; but again words struggled and failed to catch the essence. Nevertheless, J tried for the sake of my point of reference, which we both needed to maintain, and in a further figurative image he depicted himself as a child being embraced with his parents: 'If I could feel whole, I would have a relationship with my parents and feel like their child again.'

Change is never straightforward or clear: the areas we address when we look at ourselves are multi-layered, simultaneous and confused, because this is how they are layered in our sense of self. However, a further image (Figure 6.5) and accompanying words again gave us our reference point: 'fire is love, it is the glow you feel around you when you're in love, fire is pain and comfort, when I was hurt my mother would not hug or comfort me, she would give me an aspirin and put me in front of the open fire ... at the bottom right the fire seems to be setting the bird free'. I am chilled, I try not to show it or judge, and J begins to question himself out loud, 'what do I mean by that? Fire kills it does not liberate.'

J then created a ten feet by six feet image, an explosion of memory and feeling, unplanned and without order (Figure 6.6):

> Those are the hands of the children, pushing me, teasing me, that must be the school, why is the school on fire? That is my father in the bed, he is dying, I feel helpless. Those are the planes he flew in during the war, they're over the bed, over him with the guitar, I'm looking up at him, behind him his faithful wife, my mother, the woman who on many occasions said 'I only had the other twelve children to make up for my indiscretion, my eldest son, when you failed to return from the war, they said you were dead, how could I have known you were still alive.'

J appeared angry and paused, before he continued, 'I was one of the twelve, that is me inside my wife ... here are the women, the pictures, both, neither, my wife is telling me to destroy them ... what could I do? I couldn't tell where one ended and the other began ... that is the abuser, he made me set the fires ... no, I set the fires ... he did something to me twenty-five years ago, but I set the fires.'

In Figure 6.7, J returned to the day when he had tried to tell his mother about his abuse; his 'big' self talking to his 'smaller' self – reassuring, encouraging and nurturing. Magically we witness an individual attempting to change fundamentally an aspect of the internal landscape that is his personality, and we see him accepting himself and rejecting the abuse, rather than vice versa. In a final image, J returned to this sheltered

Figure 6.6 'Pictures or people ... I couldn't tell where one began and the other ended.'

Figure 6.7 'If I could go back and tell the child.'

area, where he had first been able to question his view of himself and his self-blame. He had just heard that he would be returning to prison because his offence was viewed as too serious for a medium secure setting and yet not serious enough, in view of his engagement in therapy, for higher security. In this last image, the adult and the child sit together side by side and comfort each other, which ironically reveals the greater integration of these previously separate aspects of his personality.

This case study is a snapshot of what could have become a more complex and longer-term piece of work. As part of the assessment designed to identify needs and help constructively plan J's next step, every interaction, effect and change was magnified and concentrated. J moved through many dimensions and aspects of therapy within the microcosm that is assessment.

Images can be frightening and can intrude where they are unwelcome in the same way as they can be expelled, as in this case, where they are unwanted. To see into another person's internal landscape can have effects on our own. These images and the recounting of them by the professionals, who viewed them as part of the assessment process, took on a life of their own beyond the paper and paint. They disturbed those who are not used to seeing another's fantasy and perspective on the world. Forensic clients are as capable of producing violent, disturbing image as we all are, but because of their actions or inactions the fantasy is one step nearer and sometimes indistinguishable from reality, and this can be disturbing. Art therapy has the added responsibility of being the custodian of private and sometimes dark views of the world. As a therapist, if I know a fantasy and I am able to work with it, then we are safer because it has been brought into the light. Yet, in stark contrast to this we are taught to keep these aspects of ourselves in the dark. People resent seeing what they do not wish to know, even when the originator of the image needs to reject the vision and lay it to rest. Art therapy walks a thin line between making the invisible visible and keeping it concealed. We are left, paradoxically, in a position where the ability to work with images and their makers, the greatest asset that is central to our profession, is the very thing that can seldom be seen.

In the area of forensics, the phrase 'safe and secure' is often regarded in terms of the safety and security of the outside world and the community at large. In a medium secure unit this safety works both ways, and also forms a safe and secure base for the clients to disentangle themselves from an unhelpful view of themselves and the world. From this contained chaos it is possible for clients to build new perceptions of their position and place. The locked doors do not only keep individuals in; by default they keep society out.

As an arbiter between court and client, therapy has to be seen as a forum for saying that, given this individual's life experiences and the way they see the world and themselves, then what happened was in context, though it was not right or excusable. Ultimately there are no excuses for the anti-social things we do, but there are often reasons. A change, self-sought and instigated, that challenges a previously held view of the self, the world and the individual's place within it, is more sustainable than any change based on the needs of others. Thus, this story does not have a happy ending. It was not chosen so that an art therapist could show a successful case with a quantitative outcome. The example of J has to be viewed qualitatively, as it is about the changes that happened for a person who was labelled as unchangeable, rather than about any specific outcome.

While exploring the inner landscape, which is the fabric of an individual's sense of self, we are stumbling through a private world and we will trample the plant life and will muddy the water. In truth we do not belong there; it is a world created for a solitary existence. We are honoured to be allowed glimpses of the uniqueness of the individual's make-up, which hides behind layers of learnt behaviours, devised physical presentation and schooled, controlled speech. Through these glimpses and the access they enable, we are able to witness, reflect and support the ultimate act of creativity that is the changing and remodelling of an individual's understanding of him- or herself, which is intrinsic to their frame of reference and connection with the world.

As human beings we fear the unknown and the chaos that is its corollary. Mental ill health, or the confusion, distress, anger and alternative perception of the world that is described as ill health, is all about the unknown, the chaotic and the very stuff that cannot be put into words. Art therapy, being primarily non-verbal in nature, is at home in the unordered, the nonsensical, the felt, the experienced and the feared; giving tangible sound to that which is voiceless. Art therapy is a time, a place and a way of seeing that can gain access to this process. Art therapy cannot be done to someone, but it can be engaged in with someone.

Note

1 The word insight is often used, but for my practice this is too superficial. I do not want to hear the right words strung together in a manner that appears to suggest that there is a profound recognition of right or wrong, as the most powerful changes occur in individuals without verbal connection or conscious cognitive awareness.

7 Art Therapy with Offenders On Probation

MARIAN LIEBMANN

This chapter describes work with probation clients, especially around their 'offending behaviour', that is, the illegal behaviour which resulted in their arrest, court appearance and subsequent probation order. It includes some material that has been previously featured elsewhere (Liebmann 1990, 1991, 1994). The work described was undertaken while I was with a generic probation field team, working on an outer city housing estate in a period of uncertainty and turmoil, just before the Criminal Justice Act of 1991 was finalised and became law. This act was introduced to replace expensive prison by tougher community sentences and, within this context, probation was to be seen more in terms of punishment and control than help, with offenders being taken back to court if they failed to keep stringent new standards for reporting to probation officers. The wording of the statutory purpose of supervision involved in a probation order under the Criminal Justice Act 1991, Section 8 (1), was as follows:

- to secure the rehabilitation of the offender;
- to protect the public from harm from the offender; and/or
- to prevent the offender from committing further offences.

Thus there was still some recognition of the importance of preventative and rehabilitative work. However, if art therapy was going to be acceptable within the framework of probation, it needed to do more than help offenders in a general way; to be seen as relevant it had to address 'offending behaviour' directly.

Although the examples discussed here are taken from the particular context of the probation service at that time, the methodology and the processes described are still applicable and relevant to contemporary concerns and issues about extending the uses and applications of art therapy.

Some of the particular benefits of using art therapy with offenders, in appropriate circumstances, can be summarised as follows:

1. It can be used as a means of non-verbal communication, which can be important for those who do not have a good mastery of verbal communication. This applied to a large proportion of the probation clients I saw, many of whom were just not used to talking about themselves. A related use is the expression of feelings and experiences that are 'hard to put into words'.

2. It can also be very helpful in working with those who are 'over-verbal', who often use a stream of words to say very little, sometimes to keep others at bay because they have found professionals threatening in the past.

3. Pictures can act as a bridge between art therapist and client, especially where the subject matter is too embarrassing to talk about, or has negative connotations for the client. This applies to offending behaviour, which probation clients often find embarrassing and sometimes try to minimise because it does not show them in a very good light. Drawings can enable probation clients to look at their behaviour without having to talk about it face to face first – though this may come afterwards, when art therapist and client look at the drawing together.

4. It can sometimes help people to release feelings such as anger and aggression, and can provide a safe and acceptable way of looking at and dealing with unacceptable emotions. This is particularly relevant to offenders, who are often unable to handle their own negative emotions and end up 'taking it out on others'.

5. The concreteness of the products makes it easier to develop discussion from them. Probation clients can see in front of them what is being discussed. It is also possible to look back over a series of sessions, and see patterns and note developments.

6. Active participation is required, and this can help to mobilise people. This can be important in engaging offenders, who may not be totally voluntary clients. Such potential benefits are part of the overall objective of therapy as a process of engendering favourable change that outlasts the session itself. In the case of art therapy, artwork is used in the process of bringing about such change.

Although several factors, such as the availability of suitable space or materials, can place some constraints on the process of image-making, in much of art therapy practice few formal constraints are placed on the nature of the image itself. However, in my work with offenders, I have developed the use of a comic strip format as the main initial art form to be used in the art therapy sessions. The therapeutic process with offenders centres upon the development of a reflective understanding by the offender of the reasons why he or she committed the crime: an understanding which is essential if the client is to behave differently. In this process of understanding there are key questions such as: How did this person come to commit this crime? What is the story lying behind it? How could sense be made of it? How can the probation client learn from such an understanding to bring about behavioural change? The subject for such questions is the narrative of the offence, and the externalisation and depiction of this narrative are suitably accessible through the comic strip format. Further, the comic strip has several other related advantages:

1. Comic strips are familiar to most probation clients, and for some might be their only reading/visual experience. Introducing clients to drawing comic strips can seem less daunting to them than engaging in 'art therapy', which they may never have heard of, or may associate with psychiatric hospitals.

2. For many offenders, their crime is something that 'just happened', over which they feel they have very little control. They remain unaware of any steps leading up to their offences, and see their crimes as unfortunate happenings that they could not help. Comic strips help them to slow down their thought processes and see all the stages of the offence – including those beforehand. This can help offenders see where they could act differently another time.

3. Clients' active participation means they are able to draw the offence and what happened from their own point of view. They usually also feature as the 'main character' in the comic strip story, and so begin to see themselves as actors in their lives rather than victims or bystanders.

4. Drawing comic strips for two or three offences often reveals patterns of which the probation client was unaware, and which may hold the clue to prevention of future offending.

5. The sharing and dialogue resulting from a comic strip often lead to the identification of important life issues, such as alcohol addiction, lack of self-esteem or relationship difficulties; and these can then be addressed in whatever manner seems appropriate.

The first time I used this comic strip technique with a probation client was with R, a sex offender aged forty. He had been given a two-year probation order for an offence of indecent assault on a teenage girl on a bus, in which he had put his hand up her skirt. He had committed a similar offence three years previously, for which he had completed a two-year probation order without re-offending. However, he had also offended seriously in the past as a single man.

R also attended appointments every few months with a psychiatrist, who thought his offence was a 'cry for help' because of debts. For although R was in full-time work, he was very poorly paid and his increasing family simply could not manage on the money. After finding him a debt counsellor, I undertook the task of finding out why his 'cries for help' took such an unusual and (to others) devastating form and, with his agreement, used the comic strip technique. R's depiction of the sequence of events leading up to his most recent offence is reproduced (Figure 7.1), with a frame by frame narrative as follows:

Frame 1: Early one morning, it was raining and R's car would not start.
Frame 2: He set off walking to the bus stop, over a mile away, clutching his bag of sandwiches.
Frame 3: He arrived at the bus stop to see the bus just disappearing from view (wheels at the top).

Figure 7.1 R's offence.

Frame 4: Another bus came, and R climbed on. It was empty except for one other person, a teenage girl sitting halfway down the bus.
Frame 5: R went and sat next to the girl.
Frame 6: R tried to engage the girl in conversation.
Frame 7: R put his hand over on the girl's leg up her skirt. Both people are shown looking straight ahead rather than at each other.

While R's earlier verbal account had given the impression that the incident was 'just one of those things that went a bit far' and was a chance encounter on a crowded bus on a rainy morning, the comic strip indicated otherwise. The fact that he had chosen to sit next to the only other person on the bus suggested the offence was deliberate; and in discussion of the illustrated narrative, R admitted, with some embarrassment, that this was the case. He knew he was going to commit that offence from the moment he got on the bus.

R further explained that he had been very depressed by the family debts, and when he was asked what he thought the victim must have felt, he looked totally blank, and was obviously quite unaware of her as a person. In order to explore if there was any pattern in his offending behaviour, I suggested he might do a similar comic strip of the previous offence. Figure 7.2 was the outcome with a narrative depicted as follows:

Frame 1: R stopped at a petrol station to fill up his car.
Frame 2: He drove off, past a row of houses.

Figure 7.2 R's previous offence.

Frame 3: He saw a woman neighbour waiting on her own at a bus stop, and stopped to offer her a lift. He mentioned that the woman had a 'bit of a reputation'.
Frame 4: The woman accepted the offer of the lift and got into R's car.
Frame 5: R drove away from the bus stop.
Frame 6: R put his hand over on to the woman's leg. This is the only drawing in this sequence showing a 'frontal' view.

This offence followed a similar pattern to the other in that R chose a woman he hardly knew, who was on her own, whom he did not even look at while committing the offence, while seeming totally unaware of her as a person, or how she might be feeling. He admitted that the offence in the car had given him a 'bit of a buzz', although in both cases he had made a flimsy excuse, which he knew did not really justify his action.

The use of the comic strips revealed more information in two sessions than would have been gained in months of just talking. It became clear from these pictures, and from subsequent discussion, that R had many fantasies of forbidden sex, and the offences were only the tip of the iceberg. After some practical 'first aid' work on how he could prevent any more offences, it was arranged for him to join a sex-offenders group to discuss these matters further. It was a voluntary-attendance group and he gained great benefit, becoming a stalwart member of the group. However, after he obtained some overtime work (much needed for the family debts), it was impossible for him to continue to attend.

At about this time, I had to return to the university for six months to complete the last part of my probation qualification and I heard from colleagues that R re-offended in a very similar way to before. I wondered if this had been triggered by unexpressed anger at me for leaving. As he had done well until then, he was given one last chance by the courts, with a suspended prison sentence.

By chance, I returned to the same office at the end of the six months, and R was assigned to my case-load once more. It seemed that there was a possibility that the triggers for the offences were rows, disagreements or bad feelings towards women in his life: feelings he seemed unable to express in a constructive way, or of which he was not even aware. His relationship with his wife was unnaturally full of 'sweetness and light' and yet it transpired that they never discussed their family problems. R felt it was up to him alone to sort them out.

Given that the end of the probation order was a few months away, it was not the time to start any deep therapeutic work. However, it was possible to do some couple work with R and his wife, in which they were able to be honest with each other about the problems and to realise that their marriage did not disintegrate when they did this. They learned that they could discuss their difficulties and that this helped to solve them. The hope was that there would then be no need for R to displace his feelings on to another unrelated woman. Having monitored his behaviour for some time after the end of the probation period, as far as I was aware, he did not re-offend.

A second case example illustrates the potential for the comic strip technique and its narrative form to explore some of the events giving insight into personal history. D, aged sixteen, had assaulted a girl after being asked to 'sort out a problem' by a friend. He had a background of severe violence, his father having frequently beaten his mother until she finally left him. D could hardly read or write, but had enjoyed art at his special school and was happy to draw a comic strip of his offence. What came out of this was that D was used by others to solve their problems; and that D had a serious temper problem. Just as we were about to look at these problems in more depth, D re-offended.

In the next session, D drew a comic strip of the fresh offence and described how he had accidentally spilt a can of coke on a girl's dress at a bus stop; and how, although he had apologised, she was still angry. He could not accept this and so hit her. He had felt completely blinded by anger and could not think of any other way to avoid a similar situation, except to walk home if he saw any girls at bus stops. It seemed such an extreme reaction that I also asked him to include what happened before he spilt the coke. His drawing, (Figure 7.3) depicted the frame by frame narrative as follows:

Frame 1: D (on left) and his friend from the same youth training scheme were waiting at the bus stop for their bus home after work.
Frame 2: D decided to go to the nearby shop to buy a can of coke and a chocolate bar.
Frame 3: They noticed two girls waiting at the same bus stop.
Frame 4: D's friend started making derogatory sexual comments about the two girls, such as 'Slut' and 'Whore', and then suggested, 'Let's ask them for a date.' D became drawn into this game and joined in.
Frame 5: The bus finally arrived, with quite a crowd of people waiting to get on.
Frame 6: D chucked his can of coke down on the ground, but it went on the dress of the girl behind him in the queue. She exclaimed, 'What do you think you're doing?' and D apologised. However, the girl was still angry and pushed D away.

Frame 7: D then punched the girl and she fell backwards into the flower beds.
Frame 8: Everyone else at the bus stop rushed over to see if the girl was hurt, and D ran away.
Frame 9: Another bus came, and D climbed on. He hid behind a seat in a panicky state, as he heard sirens of ambulances rushing towards the scene.

Figure 7.3 D's offence.

When asked if he could identify the factors contributing to his violence, D listed his father, violent videos, thinking about violence, and problems with girls. While both of D's assaults had been on girls, he could not think of any particular reason for this and was himself puzzled about it. D's training supervisor was of the opinion that D picked on girls because they were weaker than himself, but the prelude to the offence, as revealed by the comic strip, suggested that there was more to it than this. D and his friend had viewed quite unknown girls as 'sluts' and 'whores', which suggested a wholesale denigration of girls and women.

In further art therapy, which was used in an attempt to gain a greater understanding of his violence and its roots, D drew pictures of his parents and, in particular, of his father. Some were of good memories, such as a visit to the zoo, while others depicted bad

Figure 7.4 D's father beating up his mother.

memories, such as his father beating him up after a drinking bout. One picture showed his father beating up his mother when D was aged about ten (Figure 7.4). His mother (far left) is sad, and looks as if she is being picked up by the scruff of her neck by D's father. D (far right) and his younger brother looked on, and while his brother was crying and fearful, D felt only anger, and vowed to kill his father when he was old enough. Accordingly, we discussed whether this burning desire to avenge violence lay behind D's willingness to defend all sorts of others, getting into trouble himself, particularly as D also said the scene in Figure 7.4 sometimes came back to him when he got angry or was about to hit someone. Despite this, I sensed that D was still very fond of his father, so that we needed to find a way for him to separate his father from his violence. I also asked him whether his father had used any verbal aggression while hitting his mother, and he mentioned a whole range of insulting and filthy language, such as 'You fucking bitch!' 'Slut!' 'Whore!' 'Why don't you go back and live with your fucking mum?' The possibility that these phrases suggested to him that girls and women were worthless, seemed a new idea to him.

Some weeks later, while discussing a TV programme, D started talking about his views on girls and women. He felt they were like 'dirt on the ground that you walk on', made to be skivvies, and should not be allowed to do certain things that should be reserved for men as the dominant ones. When asked if anyone else in his family had ever said things like that, he acknowledged that his dad had. This also seemed a revelation to him.

Alongside this view of all girls and women as 'whores', D thought the world of his mother. This suggested he had a view of women polarised between the stereotypes of 'whores' and 'Madonnas'; a view which has been described in feminist literature (Welldon 1988). To help D explore

Figure 7.5 Walking away from violence.

this attitude further, I devised a way of illustrating a continuum from 'bad' to 'good' and asked him to draw small pictures of all the girls and women he knew, and then to arrange them on the table from 'most disliked' on the far left to 'most liked' on the far right. In doing this, D realised that he knew nothing about the girl he had hit at the bus stop, and put her in the middle. The next week, the exercise was repeated with boys and men. D began to see the connection between thinking all girls and women were rubbish, and finding it rather easy to hit them.

D stayed out of trouble for the rest of his supervision order. When we looked back to pick out what he had learned, he specifically mentioned his ability to see trouble looming and walk away (Figure 7.5) and that he felt he was getting on better with girls. The art therapy seemed to have helped D to externalise feelings and attitudes of which he had previously been unaware, and to learn from these. The comic strips helped him see where he could act differently, and the subsequent pictures enabled him to examine more deep-seated attitudes and change his views in several important aspects.

A final example concerns M, a very young mother, aged 20, with four children, two girls and two boys under five. Her partner B, two years older, was the father of all four but spent only part of the week with them. The family had been on the case-load of Social Services since the birth of their second child, who was in care and awaiting adoption. There were issues around neglect, ill treatment, possible sexual abuse from M's father and maybe others, interference/support from M's family, school problems, and her non-co-operation with Social Services, the health visitor and psychiatrist.

Having been given a two-year probation order for cruelty and neglect concerning her third child (which she denied), she was allocated to me, as a woman, because of the suspected sexual abuse issues. Social Services were hoping I might be able to help M talk about any sexual abuse she herself had suffered, and through this, perhaps help her become a better parent to her children. It had been agreed that I would start by making home visits to acquaint myself with the family circumstances.

During the first visit, M talked incessantly, as if any pause would give an opportunity for the expected and intolerable criticism. In these circumstances, it seemed that art therapy could provide an alternative means of communication, which could cut through the torrent of words and provide a safe mode of expression. In particular, the 'comic strip' approach might provide a way into the offences she still denied, by providing a vehicle for her to express her point of view. It would also help me to understand the sequence of events in this complicated family saga.

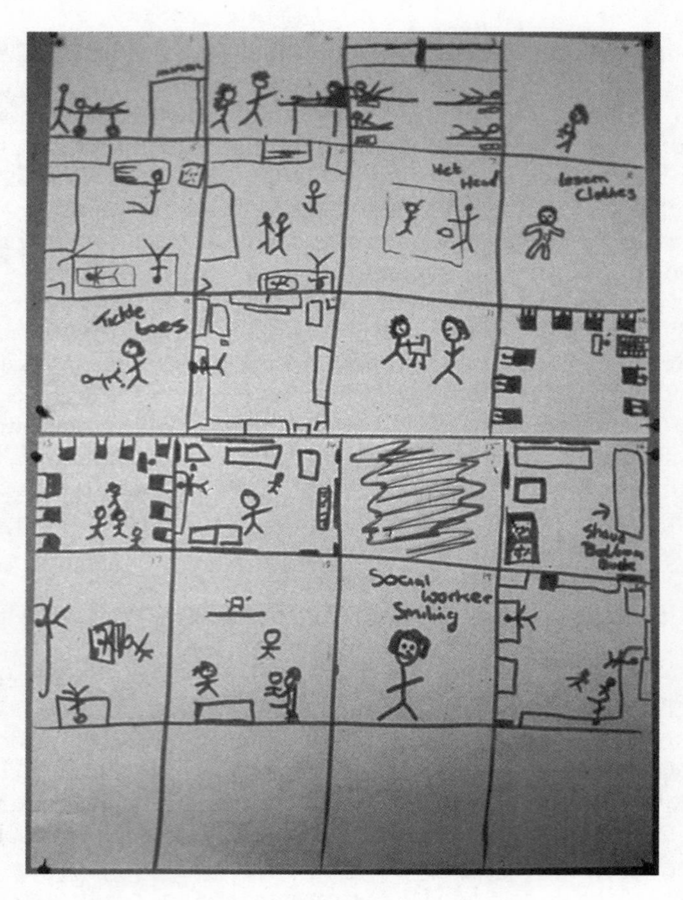

Figure 7.6 The story of L.

Accordingly we started with 'The Story of L', her older daughter (Figure 7.6), and then 'The Story of T', her younger daughter. These took several sessions, with considerable discussion along the way, with M being very hesitant about drawing and only willing to use the colour blue. In 'The Story of L', several of the frames showed a plan-like view of furniture in the rooms, and she referred to baby L as 'it', as if she had never become a person to her. The way she drew and spoke about the hospital procedures suggested she had not understood much of what was going on or the concern shown by health professionals. The penultimate frame shows the 'social worker smiling' as she took L into care. 'The Story of T' was similarly depicted.

Quite often, M would mix up her daughters' names, as if she could not separate them from each other. She also made remarks about girls never having a chance, and it did seem significant that the two girls had been more severely neglected than the boys, maybe reflecting how M felt she herself had been treated. In fact, three months after the start of the probation order, M asked for her daughter T to be taken into care as she was being 'difficult'.

Figure 7.7 M and her children.

In our sessions we moved on to pictures of each child, together with one of M (Figure 7.7). Here again the girls, both of whom seem to be encased in heavy lines, are drawn differently from the boys. For these drawings, M insisted the only colour she ever used was red.

In approaching the difficult area of sexual abuse, it was suggested that M look at childhood memories in general, first happy ones, then not-so-happy ones. She began with evident cheerfulness portraying childhood escapades, such as climbing rope ladders and trees. These were followed by less happy memories, and in one of these M used more than one colour for the first time. The scene showed their cooker blowing up, and although most of the picture was in the usual red, the flames were in red, pink, yellow and purple.

However, when we tried to look at her memories of sexual abuse, M could only get as far as drawing bathtime with her two sisters, sharing a bed with them in a holiday caravan, and feeling a lack of privacy. During the following session she said she wanted to forget any memories of sexual abuse, as remembering only made things worse. However she was happy to have a go at drawing how it had made her feel, which was like a caged animal (Figure 7.8). This figure was also done in red.

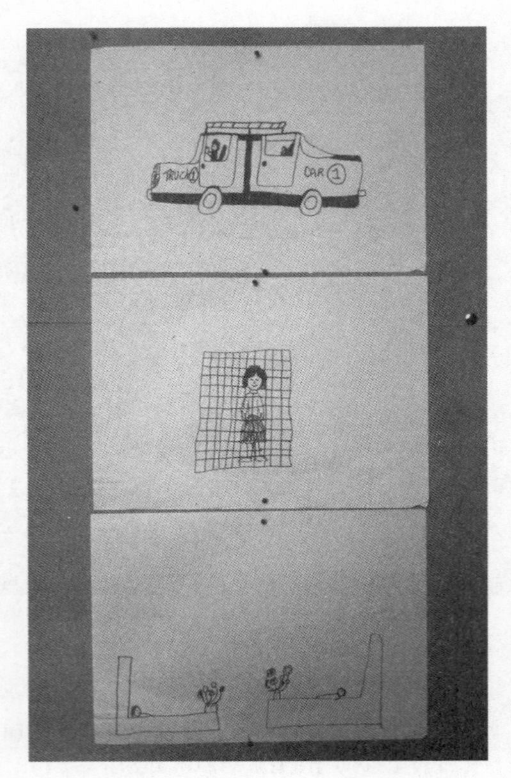

Figure 7.8 Sexual abuse – feeling caged in.

M's next memory was a strange one. It was of a bomb shelter on waste ground where she and friends played. She was happy to do a picture, and concentrated for far longer than usual, using all the colours. It depicted M and two friends looking down through a trapdoor on to a stewpot on a fire containing nine little unwanted girls. A green arm came out to grab them and they ran away. M was able to relate this to her feelings of being unwanted and of having less freedom than boys.

During these months there were frequent alarms about bruises on the remaining children (both boys), school problems with the elder, reports by psychiatrists at local and London hospitals, household debts, wardship and care proceedings. Despite all these happenings, M was always at home for our appointments, and seemed to value our sessions.

Frequently, she complained of a long catalogue of Social Services misdeeds, and these outbursts at Social Services and health professionals became the topics for several pictures. At each session she worked with great concentration on one picture for half an hour, using the full range of colours, and was relaxed and good-humoured as a result.

One picture showed a cauldron on a roaring fire boiling all the professionals she felt were interfering with her life – the welfare officer, two social workers, the health visitor, the doctor and the psychiatrist involved with her and her children. A flag at the right announced 'The mafia – a get you back soon – goody'.

Another (Figure 7.9) depicted a castle entitled 'House for kids protected from social worker and from courts and care', flying a flag with 'Protect your kids from social workers, keep them here.' The

Figure 7.9 Castle for kids.

castle's windows were heavily barred and there were battlements and a gun. The entrance was barricaded and there was no drawbridge. In the moat, sharks' fins were menacingly visible. The 'Keep Out' message was very clear.

In discussion, M acknowledged the connection between the defences saying 'Keep Out' and her feelings of being continually under threat, from Social Services and the whole world. She felt she could not trust anyone, and had no friends. Similarly, as the court case concerning wardship drew near, M drew a picture showing press photographers outside the Crown Court waiting for the result of her case. The writing said 'If you take my sons and daughter I'll get you back – Crown Courts 15, 16, 17 May.'

When the case finally came to court, a care order was made for one of her sons and a supervision order for his brother. It took a long while to find foster parents, and it was part of my role to help M cope. As the probation order drew to a close, I tried to refer her to the art therapy department at the local psychiatric outpatient department, but there were no childcare or transport facilities, and it was all too difficult.

At the time, it seemed that little had been achieved in personal terms and events had rolled inexorably on their way. It was difficult to determine if the art therapy had made a positive contribution or not, although I heard later that M had gone to the police about her father's renewed sexual advances, and had also stood up for herself when her partner had been violent to her. Also, her social worker reported that her confidence had increased, she was more open, was relating better to her children, and arrangements were in hand for her son to return home. A feeling that she herself was worthwhile was the first step towards providing that for her children. The art therapy had provided a space for this, which would not have been accomplished through words, given the extent of her verbal defences. The comic strips had allowed a non-threatening way in to an offence that was still being denied, and had given the client's perspective on events. This had led to a fuller use of colours and images, which had provided a release from the stress of continued surveillance by Social Services, and had led to the client finding her own voice.

Within the United Kingdom, there is an increasing, though perhaps still limited, recognition of the value of art and art therapy in the area of criminal behaviour, in prisons, forensic psychiatry and community settings. There are, for example, several art therapists working (mostly sessionally) in prisons, often through the prison education department. Most of those working in prisons, especially the high security ones holding long-term prisoners, use art therapy to help inmates express and counteract the destructive effects of prison life: the anonymity, the alienation, the regimentation, the infantilisation, the drab sterile environment and the dragging of time. Art therapy is seen primarily in terms of helping inmates psychologically through their sentences, although some may gain further insights, which may stand them in good stead on their release.

Some prisons have 'vulnerable prisoner units' which find art therapy useful, and some art therapists work in prison hospital wings with prisoners at risk of suicide or self-harm. A few Young Offender Institutions employ art therapists, especially for young people with special needs, and there is also an increasing number of art therapists working in Regional Secure Units and Special Hospitals, either full time or part time, with groups and individuals. These psychiatric institutions in the National Health Service cater for serious offenders with psychiatric problems. They focus far more on the offenders' behaviour, thoughts and feelings, which have brought them to offend, in the hope that something can be done to make offending less likely in the future. This is also true of art therapy at Grendon Prison, the only therapeutic prison in the UK. Some of this work is done in groups, but individual work is also undertaken.

Probation Centres, which are day centres offering short-term (usually about three months) intensive supervision of offenders in the community, may also use art therapists, although more usually they employ probation officers for the main programme concerned with 'offending behaviour' and run art, pottery, music and other creative options as opportunities to develop interests and recreational activities. However, as exemplified in the case examples discussed here, there is the potential for a more psychotherapeutic use of art therapy for those on probation, both in terms of a 'diagnostic' contribution and as a contribution to behavioural change.

In the United Kingdom, in the wake of concerns about overcrowding, poor conditions and rioting in prisons, enquiries and reports such as The Woolf Report (1991) and *Custody Care and Justice* (1991) have led to some recognition of the importance of rehabilitation. The latter report in particular recommended programmes which 'challenge sentenced prisoners about their criminal behaviour – so that they leave prison better adjusted, less likely to be bitter about their experiences, and more likely to lead constructive and law-abiding lives' (1991 section 7.2). While the reality is always less than the rhetoric, especially in times of increased use of prison accompanied by financial constraint, the Home Office has initiated some programmes for sex offenders and those convicted of violent crimes. There is a growing recognition that, if some people have to be in prison, they should be enabled to attend activities likely to reduce re-offending on release.

Several programmes using arts methods to do this are described in Liebmann (1966). It is too soon to evaluate the new Labour government's policies, but it is to be hoped that they will reduce the prison population and encourage rehabilitative regimes both in prisons and the community.

As a result of these reports, the Prison Service set up the Arts in Prisons Working Party, with an Arts Therapies sub-group looking at the most appropriate way in which art therapy can be made available in prisons. This group has written a handbook of guidelines for arts therapists working in prisons and secure settings (Teasdale 1997).

Art therapy with offenders has a considerable potential, which as yet remains to be developed. As well as the usual benefits of offering a non-verbal creative space to those in distress, it can also be used more dynamically to help offenders look at their crimes and how they can prevent future offending. As they include themselves in their pictures as actors rather than victims of circumstances, they begin to see themselves as responsible people who can take constructive steps for themselves. It is difficult work, but extremely worthwhile if it can contribute to a reduction of offending in the long term.

References

Home Office (1991) *Criminal Justice Act 1991*. London: HMSO
Home Office (1991) *Custody, Care and Justice*. Cm 1647. London: HMSO.
Liebmann, M.(1990) 'It Just Happened': Looking at Crime Events in Liebmann, M. (ed.) *Art Therapy in Practice*. London: Jessica Kingsley.
—— (1991) Letting Go and Getting Framed in *Probation Journal* (March issue) 25–31.
—— (1994) (ed.) *Art Therapy with Offenders*. London: Jessica Kingsley.
—— (1996) *Arts Approaches to Conflict*. London: Jessica Kingsley.
Teasdale, C. (1997) (ed.) *Guidelines for Arts Therapists Working in Prisons*. Croydon: H.M. Prison Service.
Welldon, E.V. (1988) *Mother, Madonna, Whore: The Idealisation and Denigration of Motherhood*. London: Free Association Books.
Woolf Report (1991) *Prison Disturbances April 1990, Report of an Inquiry by the Rt Hon. Lord Justice Woolf and His Honour Judge Stephen Tumim*. Cm 1456. London: HMSO.

8 Group Work with Adolescent Boys in a Social Service Setting

JO BISSONNET

The use of art therapy in group work with children and young people experiencing emotional and educational difficulties can be valuable in a number of contexts within social services. In this chapter such group work with adolescent boys is described, and some of the general implications of working within a social service non-health centre considered. The establishment described below is a Child and Family Resource Centre serving the whole of Norfolk and taking referrals solely from social workers.

In the United Kingdom, art therapy within the social services is not as traditionally established as in the mental health area; and as well as the restriction of stretched resources, there is a need to develop a greater acceptance of art therapy. Generally, there is an ambivalence towards the provision of therapeutic resources within social services, particularly longer-term therapy, which makes the latter difficult to argue for, even in those cases where it may prevent a re-referral at a later date. Therapeutic intervention within a social services setting, as opposed to a mental health setting, can offer a less pathological approach to children's difficulties. However the art therapist needs to be clear about how to integrate this perspective with the more traditional ways of using art therapy within a medical setting.

Good liaison between the therapist and the referring social worker is a valuable asset in a setting where the social worker and art therapist both work within the same agency. For example, specific difficulties brought up in sessions can, with the child's agreement, be discussed with the social worker. Social Service Resource Centres can be friendly and family-centred places with an atmosphere that may be harder to achieve within a hospital setting. Further, in spite of the severe financial restrictions and the continuing tension between child protection and preventative work, attempts are made to achieve a needs-led service.

The example of art therapy group work that follows arose from a number of referrals for work with adolescent boys. In establishing an effective professional context for such work within the social services, it is important that it is carefully planned and proper liaison with other involved professionals is developed. Particularly important in this respect is the process of referral and of keeping cognate workers fully informed about the setting up of groups.

In identifying a number of boys on the waiting list at the Unthank Centre to involve in the art therapy groups, I initially contacted their social workers to enquire if such would be helpful to the client. Colleagues at Unthank were also informed about the proposal in case there were individual children already attending the centre who might benefit from an art therapy group. Further, a leaflet was sent to all team managers in the areas to let the districts know about the venture. Following this, a similar leaflet was designed for the identified boys. Thus all relevant persons were clearly informed at the outset, an important process in establishing the credibility and eventual acceptance of an art therapy intervention. The text below was written for the first boys' group and sent to team managers and social workers:

> ART THERAPY GROUP
> For 11–14 year olds at the Unthank Centre, Norwich.
>
> *Aims*: To create a safe place for young people to use and experiment with art materials in order to explore and express their feelings and emotions in a visual form.
>
> Making art can also be about making choices and gaining self esteem through developing skills.
>
> It is hoped that being part of a group will help people to share experiences, find support and develop interpersonal skills.
>
> *Who can make use of the art group?*
>
> The group is for boys between the ages of eleven and fourteen who need space to develop their sense of self, and self-esteem.
>
> Those who could benefit from the group may be experiencing a range of difficulties at school, or they may have been excluded. The group could also be helpful for young people who have been accommodated because of family difficulties.
>
> *Do you have to be good at art?*
>
> Being 'good at art' is not important.
>
> Sometimes a quick scribble can mean as much as a carefully worked painting, and sometimes being 'good at art' can even get in the way of what you want to express.

> *Organisation of the group*
>
> We plan to start the group at the end of February and it will run for ten weeks. The sessions will be for two hours, on a day to be decided.

In planning the groups, consideration was given to gender relationships. It had been felt previously by the all-female team that it was not appropriate to run an adolescent boys' group with two female co-facilitators. However, the literature on this subject does not necessarily support this view.

> Male therapists can as much as female therapists run both boys' and girls' groups. It is important that male and female therapists be aware of gender specific advantages and disadvantages for both sexes. (Furniss 1991: 144)

However, the advantages of being able to offer a mixed male and female team of facilitators are as follows:

1. The male and female combination can act as representation of the child's parents, thus issues concerning the family can be addressed and explored.

2. The male worker can give the child an experience of relating to a non-threatening consistent male figure who is involved and interested.

3. The male facilitator can also give valuable feedback to his colleague on what 'growing up male' means to him and this would be vice versa in the case of a girls' group.

Another issue in planning the groups was the use of facilitators from other disciplines who may have had useful and relevant experience to contribute even though they may have limited knowledge of art therapy. The art of co-working is to acknowledge and combine different skills to form a cohesive framework for the group. It is crucial for the co-workers to get to know each other, to talk and to give and share articles that illustrate their different ways of working. In one particular group, run with a social worker with many years experience of group work in residential settings, we worked with a group of boys who had all been sexually abused. Combining our group experiences, we planned a more structured group and focused on different themes each week. The group work was cognitively based and dealt with helping the boys to understand concepts such as blame, guilt and the adults' responsibility for what had happened. The feelings the boys had around such issues were then expressed through the use of the art materials.

The theories that informed the planning and operation of the art therapy group were to some extent eclectic. First are those theories employed for understanding human behaviour. These tended to be psychodynamically based, and included Bowlby's work on separation and loss. The literature on child development is fundamental to therapeutic work with children, and Erikson's writings (1968) on the adolescent's search for identity are particularly germane to the subject. Secondly, there are theories that inform practice. A humanistic approach was used that focused on observing, reflecting back and planning session by session, according to the needs of the client group. The notion of the art-making process as a self-healing activity formed a core element in the art therapy approach to the work. As Shaun McNiff writes:

> As soon as a painting is made or a dream remembered, the images that constitute their being are experienced as wholly others. This autonomous life of the image is the foundation of a revolutionary and pragmatic treatment of our psychic diseases. (McNiff 1992: 2)

Overall, group work is a balance and overlapping between the process of the group, and the content of the tasks and activities. For example, while art therapy groups offer the chance to interact as a group, the use of art materials provides an implicit opportunity for the individual to separate and spend time alone. This process was important in mirroring experiences of separation and loss the boys had experienced in the past, the importance of which is described by Bowlby (1980).

The planned format for the groups was to use directive techniques, including drama therapy games at the beginning of the sessions, as described by Liebmann (1986), and to use the main part of the session for unstructured work. Follow-up visits were scheduled to provide an opportunity for feedback with parents and carers and to hear the boys' feelings about being in the group, without other members being present. This enabled the boys to listen to some positive feedback about themselves, and helped us to further evaluate the usefulness of the group.

An essential component of successful group work is the nature of supervision. Within the Unthank centre a system of peer group consultation operates. For every piece of work allocated to the therapist, a colleague is asked to provide supervision. The benefits of supervision are seen as follows:

1. To give workers support and a place to talk about difficult and sometimes painful issues that have emerged during sessions.

2. To discuss puzzling and complex images, play and behaviours that have been expressed by clients. To explore and plan for future sessions. Unlike managerial supervision, peer group consultation provides an equal power balance between therapist and supervisor, hopefully enabling the therapists to look at their work more freely.

3. The supervisor can contribute to the continuous evaluation, which should be part of all interventions, and forms a key element in the quality assurance of the service in general.

The ideal situation is to have supervision for one hour immediately after the group. This not only enables the facilitators to remember details more clearly, but gives an immediate space for the facilitators to off-load. Registered art therapists can be employed for this purpose, but colleagues familiar with art processes and image-making can also fulfil the role successfully.

The groups described in this chapter ran for approximately ten to twelve weeks, and lasted for one and three quarter to two hours per session. They consisted of six boys between the ages of eleven and sixteen. Some group members were outside referrals and others had previously received individual therapy at the Centre. Each group seemed to develop a style and a character of its own with the older age range of thirteen to fifteen years developing a more verbal and less active group than the younger eleven to fourteen age range. The older boys seemed to use the group in part to escape from adult pressures and demands. At these times more discussion and reflecting back was achieved.

As boys in one of the younger groups regressed as the weeks went on, activities needed to be devised for adolescents who were functioning emotionally at a much younger level. Supervision proved invaluable in these circumstances and helped us to keep focused on the boys' changing needs. In this group, meaning tended to remain embodied in the art process, and discussion and reflecting back proved quite difficult. The group developed a standard reaction against the structure offered by the facilitators. We thought about abandoning it, but decided that the boys needed to identify what they did not want to do as part of a process of discovering what they did want to do. It also gave them the opportunity to produce their own agenda as opposed to that of the adults.

All the groups developed a strong group cohesion and although, for example, there were frequent conflicts between group members in one group, they would also at times huddle together, reinforcing their group identity, during art and drama games. To help group cohesion we provided simple warm-up games and the boys seemed to enjoy the gentle physical nature of some of them. Group paintings also provided another way for the group to develop an identity. As an introduction to the groups the boys were asked to introduce themselves on paper rather than verbally, and we included ourselves in this exercise. At first they found safety by all using the same art materials simply to put their name onto paper. But as they worked in their own space they became involved in their own image-making and began to put down small but definite statements about themselves.

For example, one boy was very chaotic at home and at school, and for most of the discussion time he found great difficulty in physically keeping still. However he quickly became engrossed in the art materials and produced some creative and individual responses to the materials. He attended two of the groups and as the second group progressed, he produced as part of a group painting a disturbing image of buttocks with steam emerging from the anus. Although he had been unable to contribute to the group discussions he was able spontaneously to talk briefly about the image and link it to his past experience of abuse.

In the first group three boys produced similar masks (Plate 6). In the session the symbol of the mask was discussed with the boys, such as the 'masks' worn at school or with friends, and what feelings might lie behind these. They proceeded to spread coloured glue onto the masks, and alternated this with glitter so that a thick crust began to develop. Eventually the sticky layers on some of the masks began to drip down onto the paper below. The boy who instigated this way of working had experienced a recent traumatic bereavement, when his father, whilst on remand awaiting trial for rape, which the boy had witnessed, had hanged himself. Although he had talked about the event, he was still at the stage of numbness and showed no accompanying emotion; instead as he continued to alternate layer upon layer of glue and glitter, it seemed that these feelings became transferred as the glue fell onto the paper below. In this activity the boy attempted to construct a very thick protective mask covered in glitter, and in its slow dissolution the image had perhaps mirrored the possibility of his own mask slipping to reveal complex and vulnerable emotions. The strength and toughness of those masks that remained intact was commented on by the others in the group.

A recurring image produced in two of the groups was that of a camera (Figure 8.1). The children using this image had experienced abuse either as victim or witness. The use of the camera image seemed

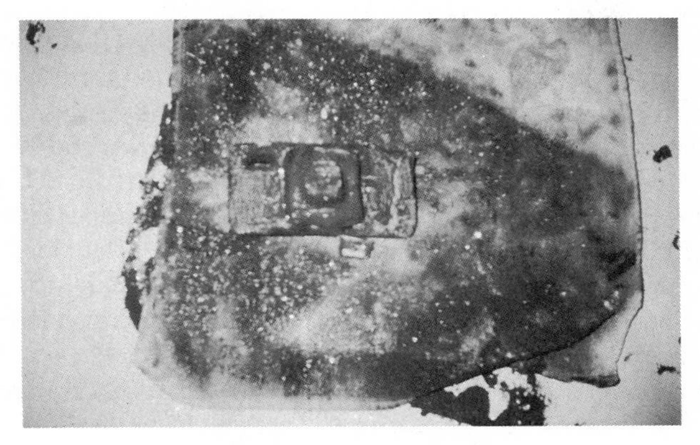

Figure 8.1 Camera image.

to express both a feeling of being exposed, and of being separated from those around them. A clay camera was made very carefully by one boy and placed on a piece of material. Black paint was then squirted onto the camera, eventually sticking the camera and material together. The boy had suffered severe physical abuse in the past and it was suspected that he had also been sexually abused. During the group he was briefly able to share this experience of abuse with two other group members, and subsequently produced the image of the camera. Thus the art-making process can act either as catharsis for uncovering and voicing of unconscious material, or as a containment and expression of feelings following discussion.

Throughout both groups, but more noticeably within the second, anxiety and excitement around sexuality was both expressed verbally and transferred onto the images. Part of this phenomenon was recognised by the facilitators as developmental, but other aspects seemed more complex. The boys expressed through images and words their anxiety about internal conflicts surrounding their sexuality. In his multi-disciplinary handbook of child sexual abuse, Tillman Furniss writes:

> Sexually abused boys are nearly always to some degree confused about their sexual identity. Fears of being homosexual are generally strong mixtures of pleasurable experience, and frightening abuse in the abusive interaction can add to the confusion. Some boys who had very frightening experiences feared they had lost all masculinity. Boys who have been aroused during homosexual abuse need to deal with the real and at times strong aspects of homosexual tendencies that are induced through abuse. (Furniss 1992: 150)

There was a feeling that for some of the boys the development of their sexuality and sexual orientation had been damaged. They had in a sense lost their freedom to choose. This dilemma was illustrated in a session where the boys began spontaneously to explore and express this part of their identity. They used a range of dressing-up clothes and chose to mix male and female dress. Boxing gloves were worn with high heels and full-length evening dress. The boy who struggled to keep still found temporary contentment dressed in a yellow sari and heeled shoes.

One of the dynamics that became apparent in the sessions, particularly in the use of art therapy, was the tension between chaos and creativity within the groups. During the life of the second group, clay was modelled and then destroyed. Other members' art products were defaced, and drama warm-up exercises would at times be sabotaged. Yet through this on-going deconstruction members also became determined to create and reconstruct. Out of chaotic behaviour, solutions were reached to salvage the situation. Chaos was depicted and contained within the group paintings. Two boys linked up to produce a painting together. They squirted and trickled paint down a long sheet of paper, working together to control the path the paint took down the paper. At first the outpouring of the paint had appeared to mirror the struggle the boys had both within the boundaries of the sessions, and internally, to contain their emotions. Through the process the boys produced a colourful and creative image. They seemed satisfied with the results, and at the end of the group the two boys carefully picked off large sections of the dried paint and requested two plastic bags to take the remnant of their painting home. One of the boys decided he would like to take the pieces home to create another picture. The process they had gone through was contained in the plastic bag, as if the feelings expressed had been made manageable.

Group paintings by their very nature are able to generate particular interpersonal dynamics around the issues of ownership and intrusion, such as over the space taken by individual members and the feelings engendered when someone's painting was painted over by another. Several boys found it difficult to acknowledge their feelings and responsibility for these actions, and the group painting generated some heated discussions. The depiction of chaos in the group paintings also expressed a developmental aspect, as if the painting had become a 'chrysalis', a juncture between childhood and adulthood. Indeed, artists too recognise this phase of chaos as part of the creative process that moulds the imagination into form. As May (1975: 140) observes, 'it is the struggle against disintegration, the struggle to bring into existence new kinds of being that give harmony and integration'.

A pertinent theme to emerge in symbolic form in the second group came about from a request to use plaster bandages. The boys chose to follow a single theme within the group and each decided to bandage their arms, helping each other and requesting help from the facilitators as well as bandaging themselves. This enabled us to speak about being hurt when the injury was not visible. Triggers for this discussion came about from the boys' comments during the bandaging, that their skin beneath the bandages felt protected and that they could show their teachers, peers and carers their bandaged arms, which would gain sympathy and recognition. The powerful symbol of the bandage gave them a way of acknowledging their own hurt and attempt at healing. Some chose to have a thick protective cover around their arm, and others much thinner. These episodes and their ensuing discussions highlighted the difficulty boys in our society have in communicating their pain to those around them.

Another important process to emerge was the expression of subculture. The most startling example of this was produced by the most silent member of the first group. A long sheet of paper had been put up on the wall and was called a graffiti space, and the boys were invited at any time to use it as they wanted. One boy in the first group took the entire space over the weeks to create a wonderful calligraphic illustration of the rock groups he most liked and admired. None of the boys invaded this space, and this assertion of youth culture became an emblem for the whole group.

The issue of gender was an ever-present theme, not only for the all-male groups but also in the mixture of the male and female co-therapists. There were times when individuals in a group became physically and verbally aggressive to each other, and though it was assumed that the male facilitator would be able to resolve this, the boys responded more positively to the intervention of the female therapist.

What does art therapy group work have to offer children and adolescents? Given that there is a lack of research as to its efficacy, my belief in the value of such work, although anecdotal, is reinforced by the home interviews with the child and carers after the group has ended, the discussion with the referring social workers and in some instances, further reports from teachers. Such reports confirm the assertions of Diane Waller (1993: 127) who includes the following in her list of curative factors of group work:

1. Observations and drawing attention to maladaptive behaviour patterns within the group.

2. The giving and sharing of information.

3. Help and support from group members when a traumatic event is told to the group and discovering that others have the same difficulty.

4. The cathartic effect of group members when a traumatic event is told to the group enabling others to share their stories.

5. Social skills and interaction, which is reflected back by group members.

6. Experiencing a safe place where thoughts and feelings can be shared with acceptance by the facilitators and sometimes by group members.

Feedback and reports both from group members I worked with and their carers seem to support these assertions, to which I would add two further points as follows:

7. For many children their only other experience of group work is in school where such children can feel isolated and rejected by staff and peers.

8. The healing power of the group creative process.

In assessing the value of group work in art therapy, it is important to acknowledge how art, as a distinct activity, was used as a healing activity in itself, as an adjunct to the psychotherapeutic dynamics, and as a support for the individual's identity. Thus art was used as:

1. Part of forming the group identity. For example the boys would at times use the same materials and produce similar images at times.

2. The art process, as opposed to the end product, was imbued with meaning for the boys. At times it was used as a focus for anger and aggression, especially in their use of clay, which would be pounded and pushed and at times thrown. Clay became a cathartic medium that enabled the boys to share their abusive experiences.

3. Art was used as a means of true creative expression, a harnessing of true creativity with healing. This was particularly vivid in some of the group paintings produced.

4. Art was also used as a means of bringing thoughts and feelings to the surface of consciousness.

5. Finally art was used to define the individual within the group: 'This is my image, and this is my mark in the group painting.'

The use of art therapy in group work enabled a number of processes to emerge, such as the group/individual dynamic, and enabled symbolic acting out to be usefully explored. The art therapy group sessions as a whole were particularly potent in enabling the adolescent boys to externalise and communicate difficult and covert feelings, whose repression appeared to be part of their behavioural problems. Boys in our society are taught to hold on to feelings, to be competitive, to hide their vulnerability. Their social interaction revolves around a world of objects and events. What then does the child do with the feelings that arise from traumatic events in his life? Girls and women may find in close friendships a safe place to express themselves. They have some command over the language of emotions. For the young male in our society there is no equivalent place.

The feedback from parents, carers, social workers, schools and the boys themselves indicated that some of the benefits of the group were as follows:

1. As part of a package of intervention for the whole family. In one case the family had attended family therapy sessions just before the start of the group. The mother, who also attended a women's art therapy group at the centre, reported that her son seemed happier in himself and was able to talk to her more easily about their difficult family history.

2. As an early therapeutic intervention after the investigation into the abuse is complete and the child is in a safe place. Parents reported in one case that their son was more confident and outgoing. The boy himself reported that it was important for him to know other children to whom this had happened.

3. As a support for children who are isolated from their peers as a result of non-school attendance. One such boy was able to make some friendships within the group and made a positive transfer to a new school.

As a boy grows to manhood he is left holding on to unresolved emotions, which can develop into feelings of anger and the need to retain power. The images produced in art therapy have an ability to seep underneath the barrier of words and behaviour. It provides a place to express feelings of vulnerability. Further, within the group process, a chink is made in the armour of invincibility that divides one boy from another. It is through this potential to repair self-perception and social confidence, that art therapy can have a valuable and distinctive role to play in the professional concerns of those working within the social services.

References

Bowlby, J. (1980) *Attachment and Loss* Vols 2 and 3. London: Penguin Books.

Erikson, E. (1968) *Identity: Youth and Crisis.* London and Boston: Faber & Faber.

Furniss, T. (1991) *The Multi-Professional Handbook of Child Sexual Abuse.* London and New York: Routledge.

Leibmann, M. (1986) *Art Therapy for Groups*. London: Routledge.

May, R. (1975) *The Courage to Create.* New York and London: W.W. Norton & Co.

McNiff, S. (1992) *Art As Medicine.* London: Piatkus.

Waller, D. (1993) *Group Interactive Art Therapy: Its Use in Training and Treatment.* London: Routledge.

9 Images of Trauma in Brief Family Art Therapy

FELICITY ALDRIDGE

Recently the study of trauma, its effects and treatment, has been gaining prominence with studies ranging from the experiences of Vietnam war veterans, to those of women who have experienced domestic violence and/or sexual abuse (Terr 1990; Early 1993). The psychological nature of trauma is defined by Johnson as follows:

> Psychological trauma occurs as a response to overwhelming personal threat in which the psychic apparatus surrenders to a situation of terror and the immediacy of death. Since the most basic psychological defences are used to preserve the survival of the self, the organization of the self is in many cases permanently altered. (Johnson 1987: 7)

This is seen by Johnson as having three main consequences:

1. Basic splitting or disassociation of the self.
2. Trauma affects our basic attachment bonds.
3. Trauma can destroy our feeling of being human.

Thus, traumatised parts of the self are cut off and suppressed by the mind and often resurface in terrifying nightmares or flashbacks. The trauma, because it is so frightening, makes the victim feel unsafe and normal bonds of attachment and reassurance are broken. The victim may try to reattach to something, and this can often be the perpetrator because of the powerful experience they both have in common. This creates very insecure attachments and the victims have a tendency to become very controlling and to avoid any situation in which they might lose control. According to Johnson, victims also have the:

> ... feeling that one's humanity has been severely compromised and that one is cut off from the community of people, that one has been

> forever soiled, marked as an outcast, or turned into a beast. The victim of trauma senses that something is gone, and only emptiness remains. To not be whole means one is no longer a member of the human race and therefore cannot be forgiven. (Johnson 1987: 7–8)

He outlines a three-stage treatment model for victims of psychological trauma, as follows:

1. The patient needs to gain access to the traumatic memories in a safe and controlled way.
2. The patient needs to work through the trauma acknowledging the events and the trauma that occurred.
3. The patient needs to rejoin the world of others.

Judith Herman in her book Trauma and Recovery (1992) also confirms these three stages, which she calls safety, remembrance and mourning, and reconnection.

Recent studies are also beginning to look at the effect of traumatic events on children. Lenore Terr (1981, 1991: 308) for example, gives eleven defined characteristics of traumatic play in children, as follows:

1. Compulsive reparation.
2. Unconscious link between play and the traumatic event.
3. Literalness of play with simple defenses only.
4. Failure to relieve anxiety.
5. Wide age range.
6. Varying lag time prior to its development.
7. Carrying power to non-traumatized youngsters.
8. Contagion of new generations of children.
9. Danger.
10. Use of doodling, talking, typing, and audio duplication as modes of repeated play.
11. The possibility of therapeutically retracing post-traumatic play to an earlier trauma.

While reading about post-traumatic stress disorder (PTSD), I became aware of how many paintings in art galleries express traumatic events, for example many paintings of myths and legends by Pre-Raphaelite painters, and more modern works such as Picasso's Guernica. The pictorial depiction of trauma would appear appropriate, as trauma is often remembered visually. This can be particularly so with children, as indicated in the work of Black and Newman (1996); Hendricks, Black and Kaplan (1993) and Terr (1990). Black (1995) for example, citing the work of Pynoos and Eth (1986), confirms that trauma is remembered visually, and that often children, while not having the words to describe what had happened, could draw a picture of the experience.

In case examples that showed post-traumatic play, Terr (1981) provides several instances of how children have used art to show the traumatic experience over and over again. Moreover, Johnson (1987: 9) regards creative therapies as having a special role in the process of healing from trauma, particularly with regard to flashbacks, of which he states: 'One of the truly remarkable facts about traumatic memories is when they erupt in flashbacks or nightmares they are often an exact replica of the event, down to every detail, as if they were photographed.' Johnson argues that art therapies have a special role in gaining access to these visual memories, as they offer a unique way in which they may come to consciousness. They are safe because the artwork is distanced from the client, and their creation gives access to remembering and mourning. Further, discussion of these images with the therapist facilitates reconnection with others.

Brett and Ostroff (1985) argue that there has been a failure to appreciate the role of imagery in clinical theories of PTSD and in the diagnosis and treatment of stressed individuals. They maintain that the use of imagery would lead to a greater coherence and focus in the study of PTSD. Stronach-Bushel (1990) and Howard (1990), two American art therapists who have written about working with art therapy and trauma, would also support this view. In discussing how psychic space has been invaded by trauma and how this can effect the transitional space in the therapy, Stronach-Bushel comments that: 'the child and the therapist form a unit capable of tolerating more affect-laden material than the child alone can tolerate; even so, the re-experiencing of the emotions can in itself constitute a trauma'. She goes on to say that 'the transitional space afforded by the art work provides a safe area in which to experiment and repeat the feelings of the trauma' (1990: 49).

Johnson, Howard and Stronach-Bushel all show in their case examples the effectiveness of the art therapy intervention. Johnson in his conclusion asks why the expressive therapies are so successful in the treatment of trauma, and wonders if art originally developed as a means of expression of, and relief from, traumatic experience. He states:

> Art, song, drama and dance in primitive times were motivated by a need for catharsis and for gaining control over threats to the community or to the individual. The arts abound at the times of nightfall, death, birth, war and natural disaster, for they help to encapsulate terror. If psychological trauma is the origin of art, is it any wonder the creative therapies hold so much promise as a reparative force? (Johnson 1987: 13)

Such studies confirm my own experience of working with children who have experienced trauma, where visual flashbacks still occur, although the trauma may have stopped. Some of these flashbacks are so strong that the children are unable to continue with what they are doing, and often at such times feel the need to self-harm. Generally, such memories of the trauma take the form of persistent visual images, often appearing in dreams. These visual memories seem to be of two types: a memory of an actual incident, and one of retribution by the perpetrator. For example, some children would see the visual image of what had happened, followed by a further image of the

Figure 9.1
(a) *top left*, father's church; (b) *top right*, basketball;
(c) *bottom left*, wrestling match; (d) *bottom right*, mother's image.

perpetrator 'coming to get them' for what the children had reported or testified about the incident.

Such persistent visual mental images and memories are disturbing for such children who are naturally hopeful that counselling and therapy sessions will stop them from recurring. However, whereas talking about the images only seems to make their occurrence even stronger, drawing and painting them helps immediately; externalising them as memories on paper and diminishing their disturbing mental presence and potency.

In the following example, a whole family became traumatised by an event; a trauma in which the use of art- and image-making significantly contributed to its resolution. A group of boys, including two brothers, aged eight and twelve, had witnessed a fight between two men outside a church, in which one of the men stabbed another to death and fled. The police were called and, as all the witnesses were under sixteen, their statements were taken in the form of a video interview. The brothers were then taken in a police car for a ride around the town to see if they could find the killer. Surprisingly, they did see the man leaving a building and subsequently he was arrested. The brothers were asked to identify him in an identity parade, which was a stressful event for them. At the time the witness statement was taken, and the children were told that if they had any worries they could come back for counselling. Two weeks later the mother rang to say the children thought that counselling would be helpful as both children were experiencing anxiety and were frightened that the murderer would 'come and get them'.

From experience I knew that if one person, or in this case two people in a family suffer trauma, then often the rest of the family will also be suffering and preventing the trauma from being resolved. The parents could not find ways to reassure the children and the trauma was so great they all believed the murderer would come, even though he was in jail. Therefore, it was decided to see them as a family and they were seen one month after the murder. It was explained to them that, as an art therapist, I found art a useful way for families to express how they were feeling.

S, the eldest, drew for his first image a picture from his favourite sport, basketball (Figure 9.1b). The words 'basketball player slamming the ball into the net' took up most of the page. The image of the player putting the ball into the net seemed to be from above and was placed in the target area that was drawn in the four colours red, black, blue and grey. The player had one arm up ready to shoot, and the other at his side. B, the youngest, made an image from his favourite sport that was wrestling and depicted a match between two characters from television (Figure 9.1c). The bigger figure, Yokozuna, was standing and his opponent (the Hulk?) lying down. Interestingly, a member of the audience was depicted as shouting 'get up please', while the standing figure was saying 'you die'. Again both the figures had one arm up and the other by their side. This picture was also drawn from the above perspective, which is often the case with trauma. There was a boundary at the edge of the paper provided by the ropes of the ring.

The mother drew an image of two pink, smiling children holding hands in a landscape where the sun was shining in a blue sky (Figure 9.1d). The figures were large and placed centrally. The father drew a picture of a church on a hill, surrounded by trees and bushes (Figure 9.1a). There were no figures in the image and the church was high up and central with a fence around it.

While the family were working on their images they were also recounting the facts of what had happened and also discussing their images. In the mother's discussion of her picture, she spoke of how she felt that her children's innocence had been taken away and how they were having to confront adult feelings about why people behave the way they do. Also she was concerned that they were having to deal with an adult system of justice that needed witnesses and facts in a world no longer that of innocent play, where the dead come back to life at the end of the game. Discussing his image, the father touched on how he felt his children were having to look at issues of right and wrong and how bad things happened that we were powerless to do anything about. The father felt this questioned the way morals were discussed in society.

These family pictures can be interpreted as containing all the elements of the murder and its effect on the family. The images by the two brothers can be interpreted symbolically as depicting the murder event, transposed into a scene composed of a visual vocabulary reflecting their own more familiar and 'safer' references, rather than those from a more literal adult world. The mother's picture contains the feelings and the father's the moral issues that the family were facing.

By the time of the next session, one month later, the family had moved house farther away from the place of the murder (although the move had been arranged before the crime). This was the subject of much of their discussion as they were enjoying the new house, especially the garden, and they chose to make an image together on this theme. This session was very informative of how the family worked together, but it also revealed how the children, while feeling safe inside their home, no longer felt safe outside. This observation led to a discussion of ways to adjust to this change, which then moved on to the theme of the loss of innocence, particularly how to adjust to knowing that bad things can and do happen. The parents felt that their own childhood had been different from that of their children, as they had not had to deal with such complex issues.

It was significant that the image of the garden was full of games, sports and other fun things. These games are also activities that would normally be played within the community, but now they were confined to the safe realm of their own garden, perhaps enabling control over who plays and participates, ensuring that no unknowns were present. The images appeared idealised, perhaps a statement that everything was all right and fine and signifying that the family were all trying to put a brave face on what had happened, as if nothing had been changed by the trauma.

Given the summer holiday period, the next session was two months later. Although the family had heard no news of how the court case was progressing, it was learned through the police that B might be called for the defence. The children were very frightened at the prospect of going to court and of the defendant being able to see them. Although the mother was absent from this session, the children and their father chose to do an image together. They liked drawing on very large pieces of paper and, as each of them was making his own image, they talked about the court case. As the session progressed, the younger boy's fear that the killer 'would come to the house to get them' became apparent. He believed that the killer would throw a knife and that he or their father would be killed. B was still very worried about retribution, and the picture started to contain other things that frightened the children. These fears, confirmed in discussion, included being afraid of each other, the fear that their mother would not buy them any forthcoming Christmas presents and fear of a wrestling clown from television who frightens children by saying that he is going to eat them.

The fears around the killer were mixed with the more ordinary fears that children naturally have, with perhaps the presence of the latter being triggered and reinforced by the trauma still exerting an influence. For example, while the images of frightening things produced by B seemed on the surface very normal and were given different names, they all looked very similar. Perhaps this was a recurring image of the killer? B's image also contained the words 'lets get him', which is what the killer said to the victim as well as what B felt the killer was going to do to him.

The next session was without the father, and the children seemed sullen. Their mother reported that they had been in trouble at school for fighting and that they had also been fighting at home. Perhaps the boys were now recreating the trauma that they had witnessed, as some children recreate trauma in the form of games to have the feeling of it again, but controlled with a reason. Trauma seems to leave a feeling or an experience in the body that is felt or remembered, and many children call this a pain that they can feel, but don't understand why. Getting into trouble and fighting often gives a reason for such pain. Trauma seems to need to be re-expressed over and over again because it is an experience that children cannot initially understand or assimilate.

Most of this session was spent looking at the information contained within the Court Witness Pack and discussing who decided who was guilty. His mother and I made it clear that several people had to agree, that B's evidence alone was not enough to convict, and that it was also the judge that ultimately would make the decision.

The last session occurred two months later, one week after the court case in which the man received life for murder. Neither of the children were called to court as their evidence was accepted in video form. All the family attended this session and everyone now felt relieved. The image of this session took the form of a talisman; a decorated box into which sticks were placed, having been decorated and written on with words that represented all the worries of the past. The box was hidden in the art room so that all these worries could be left behind. The idea for this image had arisen from conversation with the family. They wanted to use this final session to leave behind all the bad feelings of the past and move into a new phase of their lives. Responding to the suggestion that this could be done in the form of an image, the family worked together to produce something that they felt would be appropriate. The tangible image of the decorated box and its sticks felt like a positive resolution of their experience and expressed a hope that things would be different now that the legal process was over.

Altogether the sessions covered ten months, which seemed the right time span for the family. We worked together for five sessions and the brief therapy format provided the opportunity between sessions to move at the family's own pace between each of the phases involved in the resolution of trauma. The art therapy sessions enabled the family to share, recreate and change the brothers' traumatic memory of the murder scene; to externalize and objectify it so it was something that they could cope with. Clearly, as revealed in their artwork and the discussion around it, the boy's trauma had invaded and challenged the family's sense of equilibrium and stability, both experientially and morally. Through the brief family art therapy they were able to work through the three stages of trauma and recovery. Firstly, they had to work through how they could feel safe again with the experience and knowledge of the murder. Secondly, they had to be able to remember and mourn those aspects of self and family that had been lost from the time the brothers witnessed the murder. Finally, they had then to be able to reconnect together as a family. In achieving this, the artwork and image-making were paramount in externalising and controlling the disturbing memory of trauma, and in providing the opportunity and means for understanding its effect upon the whole family.

References

Black, D. (1995) Talk given as part of series in Child Psychotherapy, The Brighthelm Centre, Brighton.

Black, D. and Newman, M. (1996) Children and Domestic Violence: A Review. *Clinical Child Psychology and Psychiatry 1* (1) 79–88.

Brett, E. and Ostroff, R. (1985) Imagery and Post-traumatic Stress Disorder: An Overview. *The American Journal of Psychiatry 142* (4). 417–24.

Early, E. (1993) *The Raven's Return.* Illinois: Chiron Publications.

Hendricks, H.J., Black, D., and Kaplan, T., (1993) *When Father Kills Mother: Guiding Children Through Trauma and Grief.* London: Routledge.

Herman, J.L. (1992) *Trauma and Recovery.* New York: Basic Books.

Howard, R. (1990) Art Therapy as an Isomorphic Intervention in the Treatment of a Client with Post-Traumatic Stress Disorder. *The American Journal of Art Therapy 28* 79–86.

Johnson, D.R. (1987) The Role of the Creative Arts Therapies in the Diagnosis and Treatment of Psychological Trauma. *The Arts in Psychotherapy 14* 7–13.

Pynoos, R.S. and Eth, S. (1986) Witness to Violence: The Child Interview. *Journal of the American Academy of Child Psychiatry 25* 306–19.

Terr, L. (1981) Forbidden Games: Post-traumatic Child's Play, *Journal of American Academy of Child Psychiatry* (20) 741–60.

—— (1990) *Too Scared to Cry.* New York: Basic Books.

—— (1991) The Play Therapy and Psychic Trauma in O'Conner, J.D. (ed.) *The Play Therapy Primer.* New York: Wiley.

Stronach-Bushel, B. (1990) Trauma Children and Art. *The American Journal of Art Therapy 29* 48–52.

10 A Gestalt Art Therapy Approach to Family and Other Interpersonal Problems

JOHN BIRTCHNELL

This chapter concerns a form of art therapy developed for a particular purpose, namely getting at and working with difficulties in relating within family and other interpersonal settings. It is not the only form of art therapy that can be used for this purpose. It differs from other forms in that, while it centres on work with an adult member of the family, other forms involve work with a child (Case and Dalley 1990), and others again involve work with the entire family (Landgarten 1987). Where it focuses on the thoughts and feelings of a single individual, other forms work on the interactions of individuals within a group (Waller 1993). While it is underpinned by a belief that psychotherapy should be the primary focus of attention, and that the art should serve an adjunctive function (Birtchnell 1984), other forms are based upon the belief that the art should be the primary focus of attention, and that any therapy that may occur would be a consequence of the art process (Maclagan 1994). Whereas, in approach it involves the therapist sitting next to people while they talk and produce their pictures, others leave them to produce their pictures by themselves, to be talked about with the therapist later. While the therapy as presented here is simple and pragmatic, other forms are complex and psychoanalytical (Schaverien 1995).

The form of art therapy described here has evolved over the seventeen years I have been running a workshop at the annual Leeds Summer School (Birtchnell 1989), previously the Leeds Spring School (Brown 1980). At Leeds, four or five workshops, each with twelve members, run concurrently over a five-day period. Throughout these seventeen years each workshop adopts a different approach, but throughout I have adopted a Gestalt therapy approach that involves working with one group member at a time (Fagan and Shepard 1972; Perls 1976). Within this, I have experimented with additional techniques, and those that have worked have been retained and developed further. Whilst this way of working has evolved within the Leeds setting, it has also been used elsewhere in one-day workshops, week-end workshops, and in a weekly workshop extending over a five-week period. Although it is a personal way of working, it can be learnt by others or adopted or modified to suit particular needs.

To begin with, it is important to work in the right environment, such as a room that is soundproof, light, warm and carpeted, and free of furniture and equipment. Preferably people should not be able to see into it. It should have a ring of large cushions on the floor, and there should be a box of tissues, which can be slid silently and easily about the floor.

Because during the course of the work participants are asked to do some unusual things, it is important at the outset to convince them that I know what I am doing and that they are safe with me. I have to carry them along with me and lead them into, what is for both of us, frightening territory. I also need to convey that whatever I ask of people I am prepared to do myself. I take part in every exercise I propose. Workshop members are aware from previously circulated descriptions of what the workshop will entail. As will become clear, much of the work has to be highly directive; I need to be reassuring, encouraging and even persuasive in urging people towards, and holding them to, the mental pathway that they have to follow to reach the area of difficulty. Whilst sometimes I may suggest ideas to them, if these do not feel right, they will reject them.

Working on a carpeted floor, for hours on end, is an unusual experience. It brings everyone down to the same level. It is also a child-like thing to do and makes us all a bit like children. It seems wrong therefore to be wearing shoes, so the group members are invited, one by one, to take them off. They are also invited to take off any socks or stockings. Inviting them to do so, one by one, turns it into a ritual – everyone consciously does it in front of everyone else. It is like stepping into a swimming pool. Baring the feet is a mini-act of self-exposure. There is something important about the act of showing and being looked at. Daring to show something personal and allowing others to look at it requires a certain courage and a certain trust.

Before we start, it is preferred that nothing is known about group members apart from their first names, but somehow, very quickly, we have to gel into a group of friends who want to work together. Having fun creates a feeling of togetherness, and the first thing we do is play a silly game that helps us remember each other's names. After this we crawl about the floor giving each other a welcoming hug. Working intimately with people does not come easily either for the participants or myself. Therefore, at the beginning of the first and every subsequent session, there is some form of warming-up activity that involves touching and/or getting out of breath.

The guiding principle throughout the workshop is to find the emotion and follow it, either within one person or from person to person. The aim is to keep the level of emotion high, because emotion in one person evokes emotion in another, and so on. If the level of emotion in a group drops, it creates a loss of confidence both in me and in the group, and it may take some time to build up again. Emotion is important because where there is emotion there are problems. Emotion is to the psychotherapist what clay is to the potter. As long as the person I am working with is feeling emotional, or preferably becoming more emotional, we are on the right track. If it drops, any progress towards a therapeutic goal will stop. It may drop because we have drifted off the trail, or because that person has taken fright and has erected barricades against any further progression.

Kaufman (1996: 302) is a psychotherapist who believes that 'all experience becomes stored in memory in the form of very specific and quite separate scenes, scenes that also comprise important visual, auditory and kinaesthetic features'. Not all memories assume a visual form and music is an obvious example. However memory is much more visual than we normally appreciate, and that is why encouraging people to draw scenes is such a powerful strategy. Kaufman encourages people to visualise scenes, but what he does with these visualisations is similar to what I do with the pictures. However, psychotherapy is essentially a verbal exercise, and one cannot do psychotherapy without using words. I introduce words from the very start. They are used to give direction. Talking is encouraged and sustained throughout the therapeutic process, and although the pictures that are produced are an important component of the procedure, it is within the words that are spoken that the therapy takes place.

At the start of any group I am faced with a ring of people I know nothing about. Inside every one there are memories, some very disturbing ones. Some of these are near the surface and pressing to come out; others are more deeply buried. As a way of starting, I move around the ring relatively quickly, spending about ten to fifteen minutes with each individual, making contact and getting a feel for the kinds of issue that are around. Even during this opening round, it may happen that one person triggers off another, so when finished with someone I ask if anyone can identify with them or is in a similar situation. The person who can, or is, I move on to next. This keeps the emotion going.

I sit on the cushion next to the person and talk quite quietly. I start by suggesting that there might be some area of difficulty, involving other people, which he or she might like to talk about. Sometimes we will talk for a bit without starting to produce a picture, but at a suitable moment I will suggest that we start. The method that has worked best is that of inviting people to talk and work on their pictures at the same time. All that is asked is that the picture includes real people, rather than abstract shapes or matchstick figures, and that it is about some area of family or interpersonal difficulty. Under these circumstances, wax crayon or charcoal is preferable, but there have been successful sessions when paint has been used. If, during this reconnaissance round, people begin to show signs of emotion, I still move on and complete the round, but make a note of it and try to return to them before the emotion has subsided.

Having completed the reconnaissance round, I move on and work with someone in depth. It is important whom I start with, because if he or she brings out disturbing memories, it triggers off disturbing memories in others and the group really begins to work. If that person is defended, then others become defended and the group shuts down. The person who has the most to reveal is often the one who is the most reluctant to start, and the one who has the least to reveal is often the one who volunteers. Therefore, volunteers are not responded to unless they also convey some sense of urgency. The decision about whom to work with in depth is determined by what came up in the reconnaissance round. If someone has begun to show emotion I am inclined to start with them. Once settled down with someone, the session can last for up to two hours. If it lasts any longer we usually take a break. During this session, the two of us will be sitting close together, talking quietly; both with our eyes fixed on the paper in front of us. The rest of the group are watching and listening, usually intently, making themselves as comfortable as they can, propped against a wall or lying across some cushions.

There is a general understanding that while I am working with someone, other group members do not speak. It is normally clear, as the session continues, that it would be both difficult and undesirable for anyone to break in. Not uncommonly, other group members become affected by what is happening, and sometimes more than one person starts to cry. At these times, the group tend to move together to support each other. Towards the end of a session I invite others to join in, particularly those who have been upset. Usually, what has been explored has resonated with similar experiences in their lives. The person who has been the most affected is the obvious choice to work with next.

In this method of working, it is made clear that the picture should not be an elaborate piece of artwork, for creating good art is not the object of the exercise. It may amount only to a rough sketch, but often that sketch is absolutely right. It may be economical of line and detail, but direct and to the point. It contains only what is necessary and says all that needs to be said. Pictures that are made under these conditions (in the heat of the moment) are much more alive and powerful than those that are made as part of a suggested exercise (from cold). This is because they come straight from the person's inside, without any preparation or forethought. They just come, just as the flow of words just comes. People who claim to have no artistic ability can sometimes create amazing pictures under these conditions. They do it without realising they are doing it. People who are trained in art do not produce artistic pictures under these conditions. Their pictures do not look professional. They have the same direct simplicity.

As earlier stated, the verbal as well as the visual has a key role. At the start of the session the talking might go something like this:

> This is me ... I've got spindly legs which I try to keep covered up. I haven't got much of a bust ... I used to have long hair but I've cut it off ... I'm holding some books to show I'm a student. This is me mum ... I've put her next to me because we're very close ... She's fat but very mumsy ... She's always got this big smile ... This is Kevin my boyfriend ... He's got spiky red hair and glasses. Me mum thinks he's a nutter ... but I think she likes him really. I'm going to put me dad over here ... He's skinny like me ... He's always got a scowl. He keeps to himself a lot. We ignore him.

The next two points are controversial and sometimes strongly disputed by more orthodox art therapists. Firstly, people are allowed, even encouraged, to write words on the picture. Secondly, the picture is not to be regarded as the exclusive domain of the person being worked with, and I add to the picture myself, particularly by drawing lines and by writing words on it. This is because we are not concerned with creating a work of art as such. Producing the picture is a means to bring about, enhance and reinforce a therapeutic process. The sheet of paper is purely and simply a workplace on which we work together. When we are done, it is a mess of sketches, some drawn or painted on top of others, with words and lines scrawled over and around them; but it is often a precious mess that some people keep for years afterwards.

It is not unusual for even quite a large sheet of paper to get so cluttered up with shapes, words and lines that it becomes difficult to continue working with it. The decision to move to a new sheet should not be taken lightly. It is sometimes better to stay with a heavily worked picture because so much has gone into it. We become used to the paper because of all that the shapes and the words on it mean, and because it contains representations of the key figures in the person's life. Sometimes however, it is desirable to change to another sheet, either to change direction or to get out of a jam. For example, a woman who was working on a problem between herself and her husband began to talk of having lesbian fantasies, and it seemed appropriate to move to a new sheet to draw these. When a sheet is replaced it is pushed forward so that it is still in view and the continuity is maintained. Sometimes three, four, five, or six sheets are worked through in a session, so the entire floor area within the circle becomes covered with sheets of paper, but kept in sequence. Sometimes we move back and forth between sheets.

An important principle in this kind of work is that I should always remain a step or two behind the person who is doing the work. As in the example above, she knows where the trouble is, I do not. Only she can take us to it. I sit quietly beside her, listening attentively, responding perhaps with amusement or shock to something that she says or draws. This is important because it makes her feel I am right in there with her. It also confirms for her that it *is* amusing or shocking. I might write 'MUM', 'KEVIN', or 'DAD', under the appropriate drawings. I might also write, beside the appropriate person, some phrase that has affected me, such as – 'He keeps to himself a lot.' I always write it exactly as it was said, because what she says, and how she says it, is an integral part of the picture. There is a need to catch it and preserve it for later. I may want to bring her back to it, but at this stage, I do not want to interrupt the flow. Writing is a silent way of communicating, and writing the words means that I do not have to speak. It is better that only her voice is heard because she is the narrator. Seeing her actual words, written as she has said them, shows her that I have heard and registered them, but it also conveys that I am affected by them. Seeing them appearing in front of her increases their impact because they cause her to realise what she has said spontaneously. Sometimes this shocks her. Occasionally I have had the experience of the person writing my words on the paper. This has certainly shocked me.

I stated above that, 'she knows where the trouble is'. In considering the technique of these sessions, when perhaps it is permissible to be a little less serious, I might change this to 'she knows where the treasure is', because following someone to the source of the trouble is like trying to find buried treasure. How is it found? By trying to pick up a scent, like a bloodhound. The scent is emotion. It may be picked up from the words that are said or the way they are spoken, or from the people who are drawn or the way they are drawn. Sometimes, I might notice that the person, while talking, is drawing a little spiral in the corner of the paper, by pressing the crayon or the charcoal round and round, harder and harder into the paper. Something emotional is going on here, and I might just take note of it or switch our attention to it.

Both words and pictures carry emotion, but the emotion can be spread around in them and there may be a need to focus down upon it. When a person makes a long statement, or creates an elaborate picture, there is emotion in there somewhere, but it has to be distilled out. Short sentences are more emotional than long ones, and short words are more emotional than long words. The principle is: say what you need to say, then stop; draw what you need to draw, then stop; throw out the superfluous words, even the superfluous syllables, or the superfluous parts of pictures, and concentrate down on the ones that carry the emotion. Simple sketches are encouraged because they are more likely to catch the emotion than elaborate artworks. In a long flow of talk, there may be one sentence that carries the emotion, and in that sentence there may be one word that carries it, perhaps revealed by a tremor in the voice. If the person is asked to say the sentence again, there will be the same tremor over the same word. That word then, is the way to the treasure and must be held on to and followed. It cannot be emphasised too strongly that the aim is not to find a release for the emotion, it is to reach and explore the area of difficulty (the treasure), and the emotion simply takes me there.

The idea that the phrase or word carries the emotion and that it is important to find the right phrase or word, is an established feature of Gestalt therapy. Casriel (1971) would try out phrases or words and keep slightly changing them until he got the one that released the most emotion. He would then hold on to it or invite the person he was working with to repeat it or say it louder. Another Gestalt technique is stopping someone in mid-flow, like freezing the frame of a cine film at a point where the emotion is highest, and saying something like – 'Hold it there' or 'Stay with it'. Then hopefully, with encouragement, the person would move in closer and feel more emotion. When working with art, it is useful to encircle or underline a part of the picture and to invite the person to talk more about that, or to produce a new sheet of paper and to do a new picture that is a kind of blow-up of the encircled or underlined part. Sometimes, as Casriel did, I try writing phrases or words on the paper, and changing them around.

An extremely valuable Gestalt technique is to invite the person to address his or her remarks to whomever they should actually be directed at. This is much more emotional than telling the therapist. For example, in therapy with a couple, a wife might turn to me and say, 'He's a bully.' I say, 'Tell him.' She turns to her husband and says, 'You're a bully. You are, you're a bully.' She may then start to scream and cry. Similarly, in art therapy, a woman may be drawing her husband, and saying as she draws, 'He's a bully.' I say, 'Tell him.' She looks at me strangely. I explain, 'Look at the drawing and imagine it really is him and just talk to him.' Slowly she starts to do this and gradually she gets into the part. If this works, she may progress to screaming at the drawing, and tears will start dropping on to the paper. Perhaps, as a way of escaping from this confrontation, she may revert to talking to me, and say, 'He used to lock me in our bedroom.' I correct her by saying, '*You* used to'. She turns back to the drawing and says to it, 'You used to lock me in our bedroom', and then continues to talk to her husband about that.

One of the most fundamental principles of Gestalt therapy is the here and now principle. In the process of drawing and talking, the focus may be something that happened, or is continuing to happen, in another place. Sometimes, it also happened at another time. The object of the therapy is to recreate the then and there in the here and now. Saying 'I hate you', when, for example, a woman is looking at a drawing of her husband, brings her much more in touch with her feelings about him than by saying 'I hated him.' There are, however, two important differences between the real then and there and the created here and now. Firstly, drawing it and talking about it is safer than actually experiencing it. Secondly, recreating it in the presence of the group and myself is safer than being alone in the original situation. In the above example, adopting the here and now approach would involve my saying – 'Draw the bedroom from above. Put yourself inside and him outside. Imagine it is now, and tell him what you are feeling now.' She then begins to talk in the present tense. The whole scene feels horribly real. She is shaking and pleading with him to let her out.

Talking to the picture, particularly in the here and now, is the most powerful device I know, and once someone reaches this stage, very important things begin to happen. Talking to one particular person in a picture can go on for thirty or forty minutes. The flow of this talking can be directed by quietly prompting with spoken or written words. Thus in the example above, the wife can be pushed deeper into her feelings by my saying things like, 'Tell him what you think of him.' In the mode of Casriel, shorter or more emotive phrases or words might be suggested that focus down upon the emotion, like 'You're an animal', or 'I hate you.' If the word or phrase is right, she will scream it over and over, and then dissolve into intense sobbing.

This is what is meant by reaching the treasure, or as I sometimes call it, striking oil. If these seem glib terms, they are not meant to trivialise what is an extremely important experience. They represent a breakthrough in the therapeutic process. Usually, a bout of weeping lasts only two or three minutes, but it is very disturbing to witness, and often other group members weep in sympathy. After it, the person looks up and appears relieved. We talk quietly about what she has been through. Other group members join in. She talks about how she feels about revealing such personal experiences to the group and she is anxious about knowing what the group members think of her. They are invited to tell her and she is often surprised by their approval and admiration of her. She talks about what she might do about the situation and others say what they would do. Sometimes we end by proposing that she draws a more positive picture in which she is free or triumphant. She usually does this with great gusto, using bright colours and laughing. It is often magnificent and she feels proud of it.

Deciding which person to move on to is difficult, though crucial, as the emotion needs to be kept coming. To take someone, it needs to be apparent that emotion is pressing to come out. This requires careful judgement. If the emotion is too far down it becomes a slow and laborious process, and sometimes people become stuck with the emotion half out. They may then go away from the workshop and it may start coming out more when they get home. People do not so much opt in, as are opted in by their bodies. This opting-in process is called making a bid. The phrase 'making a bid' is appropriate because it's like an auction. Members of the group will be at different stages of doing this and it is the person who seems to be nearest to breaking point that is taken. They are not always aware that they are making a bid. It is more an unconscious than a conscious process. If the bid is not very strong it is best to ignore it. It may come on more strongly later. Sometimes the bid takes the form of someone arguing with me about how the group is run or saying that they are finding the workshop very disturbing. Once, in one group, someone on the far side of the circle appeared to be making a half-hearted bid, while at the same time I noticed that the person sitting next to me was wiggling her toes energetically. I asked her what was going on and she started to cry. That was a genuine bid, and the work with her was very productive. People can sometimes feel the bid happening, like going into labour. They come up to me before a session and say they think it's coming. They point to their bellies and say it's churning inside.

The example of the woman in the bedroom was fictitious, but comes close to what does happen in these workshops. However, Figure 10.1 is a reproduction of a picture produced in an actual workshop. As such pictures are personal, I am grateful to the person for giving me permission to publish it, particularly as the names written on it have not been changed. The picture was produced by B, a fifty-year-old woman, during a two and a half hour session. B said that she had come to the Leeds Summer School

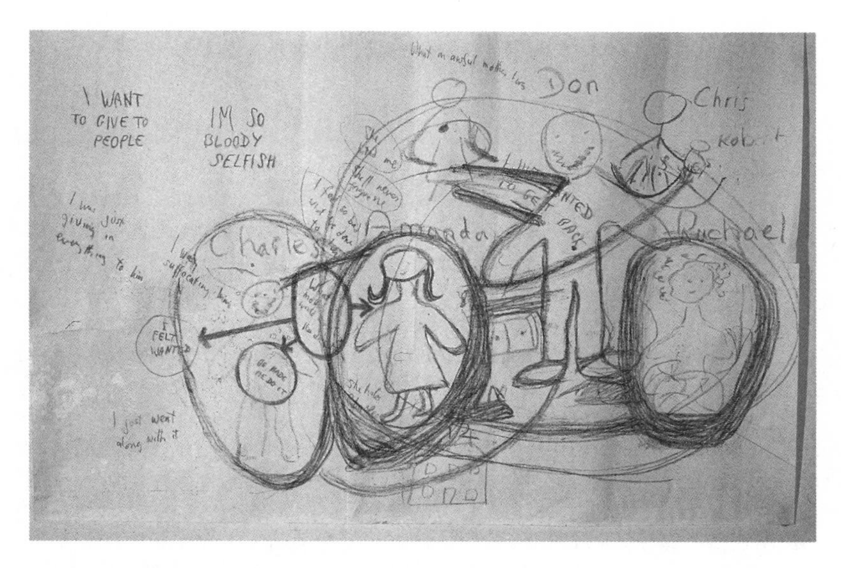

Figure 10.1 B's picture from Leeds Summer School workshop.

to learn more about art therapy and that she did not consider herself to have any serious family or interpersonal problems. She was ill-prepared for the amount of emotion that the session unlocked.

The man (Don), drawn prominently on the right side of the picture, is her former husband who now lives with a woman (Chris), slotted above his shoulder in the top right-hand corner. B herself is placed symmetrically opposite Chris above Don's other shoulder, suggesting a rivalry between the two women. Below the women, enclosed in the heavily drawn circles on either side of Don, there are her two daughters to Don, Amanda and Rachel. These are drawn with great sensitivity. To the left of Amanda is the man (Charles), with whom B went to live when she left Don. He also is encircled. Eventually, Charles left her, and since that time she has lived alone. The words written about the picture were spoken by her, and written down by me during the course of the session. The heavily drawn lines and circles did not form part of the original picture. They were created by B repeatedly drawing the crayon back and forth and round and round as she talked about the events that happened and her feeling about herself in relation to the people she had drawn.

In the session we covered three main events, each involving the release of much emotion. The first was the death of Rachel. Rachel had a congenital, progressive disease. B knew that Rachel would die one day, but it upset her to witness her becoming steadily more disabled. Rachel died at a young age on her birthday, after a party at which she ate a lot. B blamed herself for giving her too much to eat that day. She spent a long time talking to the drawing of Rachel, telling her how much she loved her and how bad she felt about overfeeding her. B thought that she had got over Rachel's death, but in the session she wept as she said goodbye to her.

The second of these events was her developing cancer of the breast on two separate occasions, resulting eventually in a mastectomy. The cancer is represented by the small black dot on the drawing of herself. Though it is only a small dot it contains an enormous amount of emotion. B had not mourned the loss of her breast or come to terms with the disfigurement, and the effect that she thought this might have had upon men. She managed to do both these things during the session.

The third event was her leaving Don and Amanda to live with Charles and Charles' subsequent leaving of her. Charles was an extremely persuasive man who was able to convince her that leaving Amanda with Don was the right thing to do. During the session, B was angry at herself for allowing Charles to influence her so, and tormented by her guilt over leaving Amanda with Don. She asked what kind of mother could do this to her daughter, and how could Amanda ever forgive her for it. She talked to the drawing of Amanda about this.

Overall the session represented a slow and gradual opening up of B's present and past life situation. Much of what she uncovered had previously been put aside and not attended to. Externalising it in the presence of the group and myself, though extremely painful for her, enabled her to face it, re-evaluate it and move forward in her life. At the end, she felt shattered but relieved, grateful and hopeful. A great deal more was achieved than would have been during a similar period of orthodox psychotherapy.

Whilst emotional release does have a therapeutic effect, it is not the primary objective of this work. Staying with the emotion, or bringing out the emotion, is predominantly a way of keeping on track. Principally, the aim is to do psychotherapy, with all that that implies. Through the drawing and the talking, the people with whom I work explore their relationships with other people. This is not an easy thing for them to do, but representing these other people on paper and talking to them does appear to be an effective way of overcoming some of the difficulties.

In this kind of work, the power of the group cannot be overstated. By the end of the workshop, almost every member of the group has had the experience of being individually worked with through a session and of being present when almost every other member of the group was being worked with. Because there are common basic problems in families, there are many cross references in the stories that are told. Members gain vicariously from both seeing each other's drawings unfold and hearing each other's stories unfold. They respond empathically to each other, and this promotes group cohesiveness and healing.

At the start of a workshop, the group members tend to fight shy of offering themselves to be worked with, but as the end approaches, there is often a rush of people opting as they realise that it is now or never. At the very end there is an important fifteen-minute terminal ritual. Everyone is invited to return to the cushion they were sitting on at the start of the very first session. This feels very strange. I then propose that we repeat the silly game we played for learning each other's names. This feels even stranger. Finally I suggest that we crawl about the floor, just like in the first session, but this time giving each other a hug and saying goodbye. Then we get up and go.

In this brief account, the methods by which a person's relating difficulties are explored within the psychotherapeutic process have had to be left out. For an account of these, see Birtchnell (1996, 1998).

References

Birtchnell, J. (1984) Art Therapy as a Form of Psychotherapy in Dalley, T. *Art as Therapy: An Introduction to the Use of Art as a Therapeutic Technique.* London: Tavistock.

—— (1989) The Leeds Experience: A Five-day Experiential, Art Therapy Workshop. *Inscape,* Autumn, 5–9.

—— (1996) *How Humans Relate: A New Interpersonal Theory.* Hove, East Sussex: Psychology Press.

—— (1998) *Relating in Psychotherapy: Application of a New Theory.* Westport, CT: Praeger (in press).

Brown, S. (1980) Dr John Birtchnell: Art as Essentially a Form of Psychotherapy. *Inscape 4* (2) 21–2.

Case, C. and Dalley, T. (eds) (1990) *Working with Children in Art Therapy.* London: Routledge.

Casriel, D.H. (1971) The Daytop Story and the Casriel Method in Blank, L., Gottsgen, G.B. and Gottsgen, M.G. (eds) *Confrontation: Encounters in Self and Interpersonal Awareness.* London: Macmillan.

Fagan, J. and Shepard, I.L. (1972) *Gestalt Therapy Now.* Harmondsworth: Penguin.

Kaufman, G. (1996) *The Psychology of Shame: Theory and Treatment of Shame-Based Syndromes.* New York: Springer Publishing Company.

Landgarten, H.B. (1987) *Family Art Psychotherapy.* New York: Brunner/Mazel.

Maclagan, D. (1994) Between the Aesthetic and the Psychological. *Inscape* (2) 49–51.

Perls, F.S. (1976) *The Gestalt Approach and Eye Witness to Therapy.* New York: Bantam Books.

Schaverien, J. (1995) *Desire and the Female Therapist.* London: Routledge.

Waller, D. (1993) *Group Interactive Art.* London: Routledge.

11 Art Therapy and the Importance of Skin when Working with Attachment Difficulties

SHEILA KNIGHT

The importance of attachment in child development has been and continues to be recognised by many researchers and clinicians, for example Brayelton and Cramer (1991), Leach (1997), and Parkes, Stevenson-Hinde and Marris (1991). Attachment theory (Bowlby 1953, 1969, 1988) encompasses many attachment patterns between parents or guardians and their children (Crittenden 1985). Therefore, it is no surprise that attachment, separation, and identity are themes that are often at the core of therapeutic work with children and their families. During my ten years of working as an art therapist with these themes, I have observed that skin, i.e. the outer surface of the body, will often be an important psychodynamic component as considered by Estha Bick (1968, 1986).

This importance of skin is perhaps not surprising, given that a child separates from its mother by the awareness of skin and that holding and handling are essential experiences for the new-born child (Winnicott 1971; Kermode 1987). Generally, the sense of touch relays valuable information regarding the environment and relationships with others. A sense of self and identity are formed by these early relationships and experiences. Attachment, security and belonging are all components of identity; and within this process, the skin is a vehicle for both attachment and separation, for coming together and for being apart.

In addition to our body skin, clothing provides an additional skin to protect and defend not only against the elements, but also from the gaze of others. Clothes as symbolic skin, through designs, shapes, colours, patterns and textures, give out strong messages that signify identity, status and circumstance. Similarly, we coat our hands, nails and faces with paints and lotions to attract attention to features such as the eyes and mouth, or to hide blemishes and scars. Body hair is also particularly important in signifying sexuality and gender. In many diverse cultures, the decoration and adornment of skin are day-to-day living experiences incorporating ancient traditions and rituals (Karlsen 1997). In Western society, the production and promotion of many beauty products for appearance and self-expression is part of a growing international industry (Coleman and Coleman 1981).

Skin, as the barrier between inside and out, has also a psychodynamic significance as the means to enable inner feeling to be externalised in the outer environment. Fluids and substances that leak out from the inside of us to the outside can be expressions of inner distress. When we are frightened the fight or flight mechanism is triggered, a component of which is the need to excrete urine and faeces. Perhaps more symbolically, there are numerous symptoms that children and families present that reinforce the importance of skin as a psychodynamic container of the self; for example cutting, bruising and eczema, soiling, wetting, ripping and cutting of clothes and so on. These symptoms can be visual markers of inner turmoil.

Similarly, if we cut or graze our skin, plasters or bandages can act as temporary skins, not only to form a seal over the wound, but to visually signify that we have been hurt and to invite emotional support in addition to physical care. The children with whom I work often request plasters or bandages to signify and express their emotional wounds. In doing so, they are using their skin to express inner emotional injuries, so that they can become visible.

The following two case examples illustrate this symbolic and psychodynamic significance of skin; firstly, where skin is used to express a confusion about identity and secondly, as an expression of attachment.

A, who presented with behaviour problems related to stealing food, is an adopted child from a different ethnic background from his adoptive parents. With the permission of his adoptive parents, and my professional colleagues who had requested my involvement and opinion as an art therapist, we agreed to meet for an assessment period. After the assessment, A and I engaged in some short-term work. During his first session A chose to create a model of a footballer (Figure 11.1 overleaf). The process took us several weeks and was painful to observe. He repeatedly made clay and plasticine clothes, the football strip, which he would then unsuccessfully attempt to glue together to create the football player. He was uninterested in suggestions of alternative techniques of how to do it. He was resolute in his attempt to create a person by joining the clothes together. The resulting figure was fragile and kept falling apart, unable to stand up or be placed in a position of action. On the outside the player looked smart and efficient, yet it had no substance and was unable to

Figure 11.1 A's striving at a model of a football player.

function. Although A was well dressed and able, his behaviour indicated that all was not well.

The process of creating the footballer continued throughout our sessions, but was punctuated with two other explorations. A liked to bury his hands and arms in the barrel that contained a sack of plaster, and when this happened his dark skin came out white. We made a plaster cast of his hand and he was adamant that the plaster was not to be coloured or painted. Therefore we had a cast of his own hand, but one where his hand was white all the way through to the centre. He then found the finger print ink and was interested to see his own unique finger pattern while we discussed police and detective work, where a person's identity can be determined by finger prints. Finger print ink is black and there is no choice of colour.

The colour of hair and skin is significant; it can mean many things, and particularly for adoptive children it is a link with their family of origin. Adopted children belong to two families, the family of origin and the adoptive family, and it seemed that A was considering his identity within these families. A used clothes, or symbolic skin, to express and externalise how uncertain he was of his self, and as in his making of the footballer, used the outer coating to show his inner hollowness. He similarly used prints of skin to try to find himself. A appeared to signify how he was unsure of his identity, not knowing who was inside his clothes. He seemed empty inside and his need to steal food appeared to be an attempt to fill the emotional gap. He appeared to have little or no information about his family of origin, and was unaware of their names. Perhaps significantly, he asked me how babies were born. A was of an age where sex education may have been a subject in the school curriculum.

Although A and his adoptive family were also receiving family work to address their difficulties, art therapy had enabled him to express, view, feel and share his experience when he was unable to voice his confusion. It seemed that during the art therapy sessions A was able to work with unconscious processes and bring them to his awareness and conscious mind. While he did not know why he stole food or why he behaved badly, art therapy had created a space where he could begin to identify and work on concerns that he had felt unable to share with his family. This in turn began to provide valuable insights to his behaviour. By being in a position to share my thoughts and observations with colleagues, they were able to incorporate A's concerns into the family sessions that primarily, but not exclusively, function at a cognitive level.[1]

B, the second example, presented various physical symptoms relating to the process of digestion and excretion. She had a history of physical investigations and operations, which had not relieved the symptom, and B experienced constipation that resulted in leakage and discomfort. She had lost her friends, and her mother and father were finding it difficult to cope with the mess and their daughter's distress. B was thus experiencing rejection. After an initial assessment period of four appointments, we engaged in long-term art therapy with regular family reviews. B and her parents also received family work to assist them with their difficulties.

After we had been working together for some considerable time, during which B had become familiar with myself and the art materials, and trust had been developed between us, she made a sculpture of a girl, and then cut her open to reveal what she termed as foul-smelling contents of excreta and vomit. Although B did not state that this was a self-portrait, it did appear to be relevant to her past experiences and her view of herself.

B made several pieces of artwork, related to a variety of concerns and then during one session she accidentally painted my hand. I did not protest, and she requested to paint my arm (Plate 7). She mixed a slimy green paint and we voiced our disgust and imagined what it might smell like: bad milk, Brussels sprouts, dog dirt, cabbage, poo and so on. As she painted my arm I informed her of what it felt like: sticky, prickly, cold and slimy. I announced that I was transformed into a monster and she squealed and began a game of chase in which I, the victim of the smearing, became the perpetrator or bully. We laughed as we played this monster game, as I declared that I was horrible and that no one would love me. B then showed me some concern and asked how I would remove the paint. She helped me clean up by stroking my arm and ladling soap and water over it. She was no longer expressing fear or rejection but was caring about my predicament.

During this session the paint had joined us together. I had become a living art object that was B's creation. B, who experienced foul-smelling liquid on her skin daily, had used my skin to express her experience and I had accepted B's projections. In Melanie Klein's concept of projective identification (Klein 1988), the child projects negative aspects of self onto the mother who contains them and enables the child to absorb these emotions back into the self in an acceptable form, and the experience that B and I were sharing corresponded to this process. I had not rejected the mess and had voiced the pain of being rejected by others. The paint had enabled us to merge by joining our skins, enabling us to become as one, i.e. I was experiencing similarities to her life experience. She was then able to take on the care-giver's role, and in effect we had joined together and exchanged roles. We were then able to separate without incurring rejection. B was able to experience an acceptance of herself, and had been able to accept the care of others.

The art therapy sessions that B and I had shared facilitated the expression of emotions utilising the transference phenomenon that had been projected and embodied via paint and skin. Art therapy had enabled B to work with primitive emotions and unconscious processes in a pre-verbal manner. Such primitive pre-verbal experience is, according to Segal (1981) a strong aspect of transference. Within art therapy the dynamics of transference and the image are explored by Schaverien (1992) who distinguishes between *embodied* and *descriptive* images, i.e. embodied refers to captured emotions and descriptive to observed emotions. Embodied images are an aspect of transference. I had become B's image and therefore embodied the emotions and transference. The transference contained primitive pre-verbal experiences, as suggested by Segal, and the process of projective identification was experienced by us during the session.

The two cases exemplify two different instances of attachment difficulties: one child was searching for self, uncertain of his identity and trying to marry two families and cultures, the other was trying to be acceptable to herself and attempting to make contact with family and friends. Both of these children clearly used skin in the process of therapy. However, many other clients work less literally and more symbolically by utilising salient skin-like characteristics of the art process itself. Skins of paint are laid onto surfaces to create images and also to protect and defend forms from the elements. Carving or etching into stone, metal, wood or lino is a process of cutting into outer skins to reveal a body of material beneath. The art objects produced by acts of creation will embrace many layers of meaning providing depth and form to feelings.

As art-making can facilitate the expression of pre-verbal and unconscious processes, art therapy has an essential role in considering how to approach attachment problems. The sensual and tactile nature of the art materials, combined with the emotional support of the therapist, encourage reparation to occur by uniting mind and body in physical acts of creativity throughout the therapeutic process.

Note

1 The family sessions referred to here were not family therapy as such, but refer to a more general way of working with families that may or may not include family therapy as derived from systemic theory.

References

Bick, E. (1968) The experience of Skin in Early Object-Relations. *International Journal of Psychoanalysis 49.*

—— (1986) Further Considerations on the Function of the Skin in Early Object Relations. *British Journal of Psychotherapy 214.*

Bowlby, J. (1953) *Child Care and the Growth of Love.* London: Pelican.

—— (1969) *Attachment and Loss.* London: Hogarth Press.

—— (1988) *A Secure Base.* London: Routledge.

Brayelton, T.B. and Cramer, B.J. (1991) *The Earliest Relationship.* London: Karnac Books.

Coleman, V. and Coleman, B. (1981) *Face Values.* London: Pan Books.

Crittenden, P.M. (1985) Maltreated Infants: Vulnerability and Resilience. *Journal of Child Psychology and Psychiatry 26* (1).

Karlsen, L. (1977) Body Ornament. Touring exhibition and talk. Oldham Museum and Art Gallery. June.

Kermode, J. (1987) A Bond For Life. *Senior Nurse* 7 (1).

Klein, M. (1988) *Love Guilt and Reparation and Other Works 1921–1945.* London: Virago Press.

Leach, P. (1997) Attachment: Facing the Professional Demands of Today's Research Findings. *Journal of Child Psychotherapy 23* (1) 5–23.

Parkes, C.M., Stevenson-Hinde, J. and Marris, P. (eds) (1991) *Attachment Across the Life Cycle.* London/ New York: Routledge.

Schaverien, J. (1982) *The Revealing Image. Analytical Art Psychotherapy in Theory and Practice.* London: Tavistock/ Routledge.

Segal, H. (1981) *The Work of Hannah Segal.* New York: Aronson.

Winnicott, D.W.(1971) *Playing and Reality.* Harmondsworth: Penguin.

12 Art Therapy in Acute Psychiatry: Brief Work

ANNIE HERSHKOWITZ

> I am aware how often I fail to convey the poetry of creative expression, the dance of liberated movement when a brushful of colour swings across the paper and the patient frees himself from his sense of helplessness ... (Rita Simon 1992: xi)

Models of brief psychotherapy, involving intensive sessions over a concentrated, relatively short period of time, have been developed from psychoanalysis in an attempt to reach a wider population. While psychoanalysis could involve several years of analysis, or be open-ended, the development of brief therapy foreshadows contemporary needs in Britain's National Health Service, where long-term therapy and treatment programmes are increasingly threatened by resource constraints. However, brief work art therapy can be effective in rapidly accessing the inner world and it is hoped the following clinical examples will illuminate this and demonstrate how art therapy is particularly important at a time of rapid patient turnover.[1]

All the examples given here were seen in the mental health services of what was a traditional psychiatric hospital, built ninety years ago. The first example is a male patient, G, aged thirty-one, who was admitted with drug-induced psychosis and suicidal ideas. He was verbally articulate, although having a flat affect and fixed gaze. He had experienced his childhood as painful and difficult since his father, whom he adored, had been frequently away on business, and G had been left alone with his mother and two older sisters who bullied him. He was referred to art therapy twice a week during a four-week period, and during therapy he was encouraged to paint, which helped him acknowledge the sources of his anguish and his difficulties with relationships. After he had finished the painting illustrated (Plate 8) parts of his childhood came back to him. He remembered he had experienced his childhood as painful and difficult, and he was able to get in touch with lost aspects of himself. It is significant, given that he had been left as a child with his critical, controlling mother and two bullying sisters, that there are three clouds above the mountain tops, three mountain peaks, and the road forks into three. Getting in touch with this lost self enabled G to work through the trauma and pain that the loss had blocked, and to integrate it and move on, instead of splitting it off and weakening the ego.

In the second example, C, a woman aged twenty-seven, had been diagnosed schizophrenic. There was no thought disorder, but she presented some paranoid features. She was pale, thin, withdrawn with poor appetite, had insomnia and chain smoked. She immediately launched into the painting of a head (Figure 12.1). The head appears about to burst forth from the page and the powerful primitive feelings can hardly be contained within the page, as was the case with the session itself. She produced a number of similar heads in succession, and then in the fifth session drew a self-portrait of her body. She erased the already slender body as soon as it was finished, and redrew it making it even more slender (Figure 12.2). The slender figure, the position of the hands and arms, and the absence of genitals suggested narcissistic regression. When I reflected on how vulnerable she appeared, she cried and said she felt as 'vulnerable as a leaf in the wind'. She went on crying like a baby with contortions of the face, mouth wide open, and with no attempt to control the increasingly violent sobbing.

Figure 12.1
C's powerful head painting.

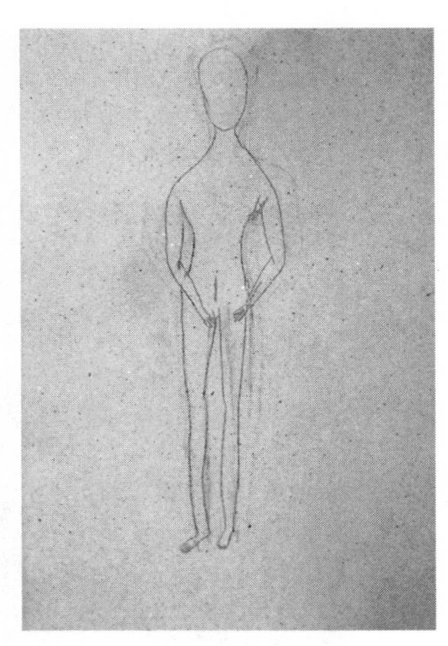

Figure 12.2
C's body self-portrait.

The painting of the first head, placed next to the slender figure, together present a graphic illustration of her mental state: how can such a frail person contain so many powerful and explosive feelings? The painting helped her release some of the emotions and provided a space in which to express them creatively. She was then able to work through her pain in a safe way. Her diagnosis was reassessed on the basis of these observations. She was discharged shortly afterwards, and was then followed up at the Day Hospital, where she continued to make good recovery.

The next example, R, a 47-year-old woman, was admitted after a manic depressive breakdown. She was childless and had recently had a hysterectomy. She attended the art therapy group twice weekly during her six-month admission. Quite early on in the treatment she painted a delicate emotional image of a mermaid. This was then followed by several paintings of coloured stripes and patterns. Then another charged image was produced (Figure 12.3) depicting a dead baby in its christening dress, shooting up to the stars from what appears to be a volcano or a womb. The scene looks like a birth gone very wrong. The mood of the painting was one of rage and violence, which contrasted sharply with the patient's gentle manner and soft voice. The patient never talked about the paintings; instead she chatted casually about daily matters. Yet she was able, through art, to express and work through her loss and grief around her childlessness, something that she could not articulate verbally. It was also interesting to see how an emotionally charged painting would be followed by a neutral painting and other abstract shapes, suggesting that some period of recovery was necessary before she was able to produce the next emotionally charged image.

Figure 12.3 Dead baby shooting up to stars.

The way in which a created image can break through verbal and behavioural defence mechanisms is dramatically shown in the next example. O, a woman aged fifty-one, following bereavement and ensuing guilt, had developed obsessional rituals, including compulsive hand washing and teeth brushing to the point of gum erosion. She was referred to art therapy *in extremis*, and was seen twice weekly for three weeks. For the first two sessions O was agitated and spoke incessantly, with pressured speech. She would not paint during sessions, but instead, she brought along knitting, done outside the sessions, to show me, thus attempting to control the session. In the third session, she was given a lump of clay to roll in her hands. For the first time she was silent, and there was a sense of stillness and centring. She produced the torso as shown (Figure 12.4).

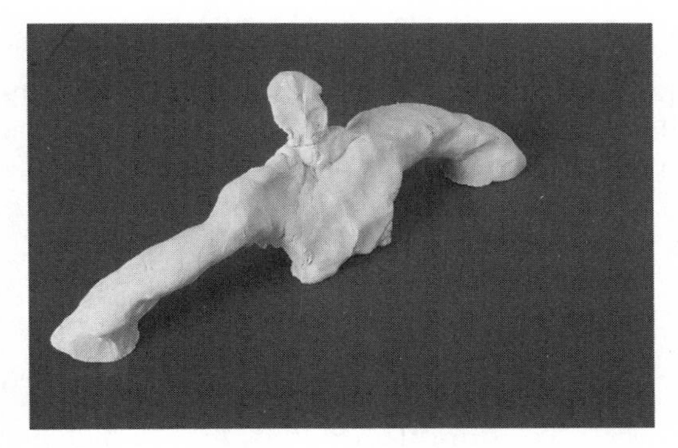

Figure 12.4 O's three-dimensional torso.

Particularly significant, given that this patient could not remove her hands from running water until rescued by a nurse, is the massive bulk at the tip of the arms, as if struggling to lift off the substrate. Also noteworthy is the cavity at the back of the neck and the disproportionately small head, indicative of her low self-esteem. The head became progressively larger in later torsos until in one following session the head itself was produced. By that stage she was crying, in touch with the feelings of loss and grief that she had previously repressed. In the next session (the fourth, and unknown to us at the time the last), she produced the bowl/ container as illustrated (Figure 12.5 overleaf). The bowl expressed the containment she experienced in the art therapy through which she was enabled to experience the feelings of pain, grief and loss, and to confront and express them. She was then able, for the first time since her bereavement, to freely talk about the loss, and to look to the future alone.

Figure 12.5 O's bowl/container object.

The use of clay, in the above example, illustrates how different media are sometimes more appropriate and effective for particular cases. Unable to lose her inhibitions with painting, the clay through its tactile physical nature was more correlative to the physical nature of her symptoms and their dynamic. The form and nature of the art work and its process can be as significant as the content of the imagery produced, particularly where the specific creative process is formally close to the behavioural dynamics and their causes.

In another example, a female patient aged thrity-one had a history of suicide attempts and self-mutilation from the age of sixteen. She could not paint, and had never been able to play as a child. She suffered from depression, felt numb, and was unable to experience emotions, although able to identify them in others. Later in therapy, it was revealed that she had been sexually abused by her father between the ages of six and thirteen. She attended twice-weekly art therapy groups on the ward during her fortnight's admission, and continued to attend individual weekly sessions as an out-patient for twenty months after discharge. At first, she was unable to use art materials, instead she used biro and ruler to produce geometric patterns. A year into her therapy, she produced thick paintings of grids which resembled the razor blade cuts on her body. Eventually, she was encouraged to work on a large scale, using paper from a 6ft (2 metre) roll, such that she had to work on the floor, on all fours, stretching across the picture. She was soon able to represent large feathered birds, of which she had a phobia, and images of her body, which she hated.

She alternated between painting and clay work. Clay work enabled her to work through deep body trauma and the complex and powerful feelings of sexual abuse. The three-dimensionality of clay afforded better fusional possibilities, while painting, on the other hand, helped mediate the more symbolic functions, gradually replacing the body and its mutilation as the medium of expression, and for containment. Different media, their scale and their ensuing physical processes had thus made different kinds of contributions within the therapeutic development of the patient.

This patient presented the typical dissociative states of the adult survivor of child sexual abuse. The splitting of affect and feelings enables the individual to erect the semblance of a functioning, adaptive self around the core of a wounded, abandoned child still searching for acknowledgement, validation and compensation. There is no vocabulary to express the experience. The amorphousness of the clay, and the thick impasto of the paint applied literally with the fingers, provided channels for tapping the rage, the chaos and the horror.

The last example is the case of a male patient, R, aged thirty-four, who was very verbal and articulate, but with flat affect and poor eye contact. He had lost contact with his father who had left home when he was eleven. Soon his mother married another man who fathered more children. R had always felt marginal, unable to experience emotions. A postman, on night shift work, he was gradually becoming estranged from his home and family (he had two sons, nine and eleven). He eventually left home and had become profoundly depressed and suicidal. It was significant that his own son was at the time eleven, the same age as when R's father left home.

We knew approximately how long he would be in hospital and this enabled the sessions and their ending to be planned, unlike other instances when discharge happens at very short notice and without consultation. We met twice a week for three weeks for individual sessions. From the very first session, we talked about endings, separation and loss, and feelings of grief and anger. R knew he should be feeling grief and anger, but he felt only numb. However, I felt very tearful during sessions, so he must have projected these aspects into me. He could not paint, and it seemed he felt too threatened by the loss of control and surrender that painting involved. He was introduced to clay, and he plunged into the bin and took out a large chunk, which he rapidly moulded and shaped with vigorous hand movements. As soon as a shape came into being, he knocked it into a new one. At the end of the session, the clay was thrown back into the bin.

We prepared for the end of treatment as the date of discharge drew nearer. In the last session, he produced an object shaped like a series of bridges (Figure 12.6). He was tearful as he handed it to me for firing,

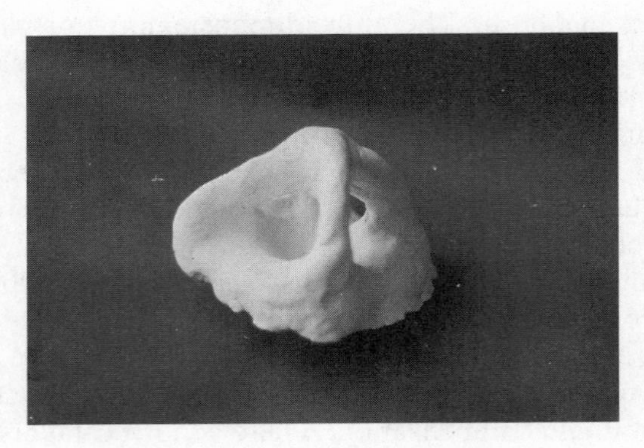

Figure 12.6 R's farewell object.

but he could now 'hold on' to an object as a finished form that he had completed: a sense of boundaries, beginnings and endings, a start and a finish, and of separation had been achieved. R had been traumatised by the loss of his father at a formative stage of his development. The loss had never been talked about or worked through. It jeopardised his own fatherhood. It appears that the medium of clay enabled him to work through his object relations, providing him with a safe space in which to explore his relationships. He was then able to work through his loss and his grief, and to start to mourn his father. He was followed up at Day Hospital, and was discharged in due course.

As all these examples illustrate, art therapy with its triadic transference possibilities provides unique opportunities for integration and for 'working through' in brief art psychotherapy. A more focused directive approach is required for short-term work where different media and their formal characteristics are important. Clay, for example, offers fusional possibilities unequalled by other media, relating more to narcissistic, pre-differentiation aspects, while it appears that painting assumes a pre-existing level of differentiation in the realm of transitional phenomena (Winnicott 1971). However, it is also significant to note that many patients use paint in a three-dimensional way, exploring the plastic/tactile/skin qualities of paint; reminiscent of clay and dynamically functioning in the same way as the more obvious three-dimensional medium.

The capacity to symbolise depends on maturation and ego growth. Initially, the infant experiences instinct driven images or archetypes. We can only experience images symbolically when enough ego consciousness has emerged; when we can relate to the paradox of separateness and identity. Symbolisation means to be able to experience the existence of links between objects that are also recognised to be separate and distinct. As Jung (1931) maintained, symbolisation is a process that involves an 'as if 'attitude. Hence the attitude to, rather than the nature of, mental contents determines the ability to symbolise. Painting, by its nature, implies a translation from a three-dimensional world into a two-dimensional one, in which the 'as if' attitude is central. On the other hand, clay mediates the archetypal, pre-differentiation narcissistic phase of development.

Brief intensive art therapy work can be effective for acute psychiatric patients by providing the dynamics for hitherto inhibited expression, and enabling patients to confront repressed experiences and feeling. Not only is such intensive work economical, but it can often reveal the important psychodynamics of a case more effectively, and in a shorter time than more conventional therapies. In this, the formal characteristics of the art media and its processes, and not just the content, are a key factor.

Note

1 Freud had already acknowledged the need for brief psychotherapy in the aftermath of the First World War, in an attempt to make it available to a wider population. This involved a more active role of the therapist, and setting a time limit for treatment. Hence a more focused approach developed. By brief therapy it is meant any number of sessions from four to twenty, but in some examples here, this is extended, where the level of damage of the individual is a factor.

References

Jung, C.J. (1931) The Aims of Psychotherapy. *Collected Works Volume 16.* 46–51, London: Routledge and Kegan Paul.

Simon, R. (1992) *The Symbolism of Style.* London and New York: Routledge.

Winnicott, D.W. (1971) *Playing and Reality.* London and New York: Tavistock.

13 Image, Charisma, Groups and Abuse

JOHN HENZELL

The Nine O'clock Service, or NOS as it came to be known, was started in Sheffield at an evangelical Anglican church in the mid-1980s by Chris Brain, later to become the so-called 'rave vicar', who had a background in rock music.[1] His idea was to attract young people to Christian worship by staging weekly 'rave'-type services characterised by powerfully amplified acoustics, and accompanied by exotic lighting and visual effects. These proved an enormous success, and he soon recruited an inner group of helpers, mostly young people in their late adolescence and early twenties. They worked largely voluntarily, existing only on small allowances. Some of this inner circle were highly talented people who donated their creative skills to the design and management of the rave services as well as other NOS activities. The inner circle was also strikingly characterised by a group of attractive and strikingly attired young women whose task was to attend to Chris Brain's personal needs; according to Howard, they were Brain's 'postmodern nuns' who formed the 'homemaking' team (Howard 1996: 69). He had a gift for PR and enlisted the backing of important members of the Anglican hierarchy, including the Archbishop of Canterbury. In the face of falling attendance the church was delighted by the great success of the Nine O'clock Service. Brain and his other NOS workers had obtained additional support in the form of equipment, money and property. There were also ambitious plans to export NOS to the United States and Brain, together with a few other NOS members, had spent some time there selling this idea. It felt to many of them that they had been chosen for an extraordinary mission. There was great excitement about all this and everyone wanted to be in on the action. Not noticing the hidden side of Brain's private relationships with those in his inner circle, the church accelerated his ordination into the priesthood. He was extraordinarily vain about his appearance as a priest, taking great pleasure in his long cassock and wearing his hair long to give a Jesus-like impression.

In the summer of 1995, cracks began to appear in this apparent success story. Allegations of sexual abuse were made to senior church officials in Sheffield by female members of the NOS inner circle.

According to these women, Brain's personal needs had been physical as well as spiritual. As well as this, many members of NOS felt both used and abused in other ways. The church struggled to accept this, but finally realised its trust had been betrayed and that it had unwittingly spawned a cult. By late August 1995 this had become public knowledge and over the next days NOS fell apart. Some members who were legal trustees of NOS's assets sequestered most of these so that Brain no longer had access to them. Alerted to the prospect of a major scandal, the media closed in with their cheque books and most of the group went into temporary hiding to escape this unwelcome attention, either in a series of hotels and private houses, or at Whirlow Grange in Sheffield, a large house and conference centre owned by the diocese.

Aware of the potentially devastating emotional effects such events might have on these people, and of consequent tragedy, the diocesan authorities and the Nine O'clock Trust sanctioned finances to provide therapy and counselling for those in need of it. In the following months this was co-ordinated by an organisation called SHARE, a group of psychotherapists with Christian affiliations – there is a strong tradition in Sheffield of links between the church and psychotherapy. Two of those most active in organising the therapeutic handling of events during the immediate crisis were Stephen Lowe, the Archdeacon of Sheffield, and Christine Nauen, a psychotherapist who had previously worked with me on the art therapy course at the University of Sheffield. She asked me if I would be willing to work with some of the victims of this catastrophe.

First of all, some disclaimers. I am not a Christian or a member of any religion. I would describe myself as an agnostic who recognises in his secular way that certain human experiences do have a religious or numinous quality to them. This is true of my own experience. Secondly, I am not an expert in the kind of events that took place within NOS. Before I became involved in this work in Sheffield I had only read about religious cults in the newspapers. This was equally true of all of us involved in this therapeutic work. Thirdly, what I am about to say is not an account of art therapy as such, though it does arise from powerful images in my clients' experience. When first meeting someone in distress it seems to me tactless to push a particular technique – rather listen and wait to see how the client wants to work. With one of my NOS clients the offer to make actual images was made, but she preferred to talk.

The latter point does raise a crucial issue. The images that subsume a therapist's work, whether they be art therapists or psychotherapists, cast their metaphorical net far wider and deeper than any concrete origin in an actual visual image might suggest. In my lengthy experience of art being used in therapy and psychiatry, I have often seen pictorial imagery used in ways that completely lack imagination; on the other hand I have worked with those whose medium was words, but who were able to invest them with the sort of transforming imagination that can change peoples' lives. This is psychotherapy's power, whatever the expressive preferences of either therapist or client – and art therapy is a psychotherapy, in the generic rather than technical sense of that term. A therapist rooted in one expressive or communicative mode may well miss what the client may show or tell in a different way. His or her style in conveying this is as much their choice as the therapist's. Once the client embarks on this enterprise, often set about with all sorts of risks and fears, I don't think what they choose to do will be lacking in imagery that consists in much more than its being literally visible. It may be however, that if therapists have a background in art and are practiced in it, they will be more than usually receptive to the metaphors and meanings awakened by these images, as well as in actual images made in psychotherapy. In this context, this chapter is about the images that surround and lie behind the lives of people, in this case two people traumatised by an abusive group, rather than about a particular clinical technique.

The wish to keep the identities of the two following clients concealed will restrict this account to some of the general principles involved, and I will be sparing with detail. It is also important to state that the widespread allegations of sexual and other forms of abuse made against Chris Brain have not as yet been tested in court; at present they remain allegations.

I first met two of my clients in August 1995. They were a couple in their early thirties, and had been involved in NOS for ten years. Most of my comments will be based on what was told me by this couple who became my regular clients. I saw them on a weekly basis for some months with a break at the end of the year. On all but one occasion I met with them individually, as this was their wish. We had our last meeting the following January, with the understanding that they could contact me again if they so desired.

What I heard, therefore, were two accounts of how they had each become sucked into a baffling and almost hypnotic relationship with Chris Brain and NOS. They were both central figures in the NOS inner circle. In important respects their narratives are told from the points of view of a woman and a man and show how the seductiveness of the situation engulfed each of them in different but subtly related ways, and affected men as well as women. On one point they were in complete agreement: a significant feature of Chris Brain's hold over the entire group was to instigate intense rivalry among them, including driving a wedge between couples such as my two clients. This served to make every individual's primary relationship one-to-one with the group's leader, rather than with peers, friends, partners, or the group as a whole.

What follows, while based on some things actually said to me by my clients, goes beyond them to indicate some general experiences of many members of the NOS inner circle, gleaned from their and others' comments. It is a kind of collage that depicts some typical events that took place within the NOS group, and while not all of them happened to my two clients, they were all first-hand experience for different people involved at various times.

First of all a woman's experience. She is in intense rivalry with her female peers in the inner circle. This rivalry pivots around the attention paid to her by the male leader. The effect of this is magical and seems to redeem all her imperfect and ordinary experiences with other men, particularly her father, but also to some extent with her partner. His ability to do and say things that seem to touch directly on her most intimate thoughts and emotions is dazzling. The leader's attentions towards her are strictly rationed. Sometimes, in the course of their daily business, he might take her to one side and say how important she is to him and to the whole enterprise. From time to time, at night, she will be rung and asked to put the leader to bed. Whatever she may be doing it is her duty to respond immediately. This may involve nakedness and her giving him massage. The situation is highly ambiguous: while actual sex rarely seems to occur, and he shows few signs of obvious erotic arousal, she is invited by him to do whatever she wishes. If they are staying in a hotel she may have left her partner in a nearby room. She will also often be aware of the proximity of the leader's most favoured female confidante and adviser amongst the inner circle; often it is she who may have rung her. Although Brain was married and living with his wife, this was someone else. During such encounters she will be told how extraordinary she is, how she understands him so well, and how marvellously soothing her touch is. To her consternation she may be sharply criticised a few days later, told she is a whore, must learn to think the right things, and practise visualisation exercises that will help her to have a correct attitude to their important work together. Outside personal commitments to her partner, her family, even having children of her own, must be put on hold in the interests of the group's world-saving mission. Her feelings towards others, her friendships and her relationships, are criticised. Sometimes she will go on shopping expeditions to buy clothes with other women in the group, under the supervision of the leader who accompanies them, telling them what is to be bought. She spends much of her time feeling miserably worthless, putting most of her personal ambitions to the back of her mind, desperately waiting for the next time the leader will single her out for his attention. Though he tells her she must confide her thoughts and feelings in him, she feels it part of her duty not to interrupt or disturb him too much.

A man's experience might be as follows. He is invited into the inner circle and given important work to do. He feels he is indispensable to all of their work and the logistics of the rave services depend very much on his skills. These may be to do with design, management, finance or planning. From time to time, like the woman, he is told how extraordinary he his, that he has been hand-picked by the leader because of his special gifts. Like her, he feels really chosen. Also like her, he feels many of his problematic past relationships are made good by his contacts with the leader. At last he has met someone who lacks the faults of his parents, and because of this he feels hopeful that the past may at last be mended. He takes pride in his work and feels some triumph in Nine O'clock Service's great successes (it was indeed superbly produced). At times, though, he is disconcerted to find himself being publicly berated by the leader. Sometimes things went wrong, maybe because he did what the leader told him to do, and although at the time he realised it was bad advice, now he has to take the blame. On the other hand, when an idea of his has worked well, the leader appropriates it as his own. He is told he can take any of his private problems and worries to the leader; however, when he does he is ignored, or grotesque tricks are played on him, and he feels intensely humiliated. He is shocked to find the leader doing things in public that he is sure refer, like a charade, to some of his most private experiences, like a subtle form of torture by suggestion. The leader might then claim these experiences as his own. At times like this, he feels the leader is a kind of mind thief, but he keeps this to himself. He also feels ashamed of the way in which, to protect himself, he colluded with the leader in humiliating others in these ways – indeed one of the functions of the inner circle was to furnish a regular supply of scapegoats. He must also put his own personal projects on hold, postpone his higher education, having a family, and not feel jealous when he knows his partner is in the leader's bedroom. Like his female counterpart, he too feels the leader has an unerring ability, a kind of psychological radar, to seek out significant parts of his inner life and use these to goad him, through praise and blame, ever further under the leader's spell.

When everything came apart at the seams both the woman and the man felt betrayed, humiliated and deeply ashamed. To use an analogy, it was as if a group of astronauts who had spent years training for a mission, and who had finally felt their spacecraft blast off and enter orbit, suddenly discovered they were the victims of an elaborate hoax and that all the time they had only been on earth in an artfully contrived film set. To be sure, they were furious with those who had set them up in this way, indeed so angry they were worried about what they might do, but, nevertheless, they still blamed themselves, as if it was partly their doing. Because at many levels personal relationships within the inner circle were deeply flawed, the real creative achievements of the NOS episode tended to be forgotten, perhaps because the leader had falsely appropriated these for himself. It had all become poisoned by him. They were also sickened when, as the scandal broke and their whole way of life for the past ten years began to disintegrate in a matter of a few days, they saw their leader trying to make out that he was Christ himself being crucified – he even held his arms out that way! But by then, his charismatic image was dispelled.

How did he manage to do all this and to take things so far before the 'hallucination' was seen for what it was? While I cannot hope to give anything near a comprehensive answer, certain things do stand out. One of them is how powerful and persuasive images can be, not literal images in the sense of their being just visual, but rather images formed from the imagination and made up from the countless nuances of human relationships. Brain certainly seems to have had an uncanny intuition for such images, an ability to locate them exactly in others when he was alone with them, and, conversely, to project his own images into others and blend them precisely with their own inner imagery. In psychotherapeutic language, we might call this an unusual species of projective identification. It was as if he were conducting a gigantic psychotherapy group in which, by an extraordinary reversal of roles and ethical responsibilities, the group leader's own countertransference became the central phenomenon. This cannot be accomplished by painstaking cognitive work, it requires the deft manipulation of deeply laid images, those kinds of images in which we store our basic experiences of living that Jungians call 'archetypal'. This is what Bachelard means when he says: 'But the image has touched the depths before it stirs the surface' (1969: xix). To Bachelard this is the image's creative function. I would argue it is also its perversity. Several people have spoken of this, and not necessarily just those involved in NOS, 'be careful when you're alone with him' they say about Brain, 'he can make you feel odd things, sympathies and emotions you didn't think you had'. Such psychological conjuring must depend upon a subtle and highly invasive aesthetics, an aesthetics of gesture, tone, atmosphere and appearances.

The Nine O'clock Service itself was a new ritual made up of images, images that were expressly meant to be hypnotically stirring and to supplant the older, some would say faded, imagery of traditional Christian worship. It was Brain's power, perhaps genius, to see that old and new images might be combined in a heady mix. He might also have unconsciously guessed that such a public experiment was ideally suited to his own implicit agenda – to play the Messiah himself. To do this, a religious institution that actually believes in a Messiah was the best possible place for him to conduct such an experiment in self-aggrandisement. And within the Christian religion there are indeed many institutions where women marry Christ, men serve God, and both relinquish personal ambitions, all in order to save the world. The difference is that these are regulated as well as segregated, and sex and personal profiteering do not take place – or at least they ought not! Indeed NOS *ought* to have been properly regulated by the church, but then it too was enthralled.

Finally, how should one work with survivors of this kind of abuse? While there are many differing views about this and corresponding practical approaches, the following would seem appropriate. Firstly, some of the reasons for not rushing in with one's own specialist therapeutic techniques have already been hinted at, and it is important to let those you see find the most appropriate language in which to begin to convey their feelings. In the case of the couple I saw most, this consisted in talking, in a repeated going over and over of events, until a whole narrative emerged. One person, whom I saw briefly, did indeed show me paintings and drawings and I am sure, had she wished, these would have played a large part in any ensuing sessions. Nevertheless, if we listen carefully we begin to see the images and metaphors upon which speech floats. This indeed is part of the art in therapy. Each of the two I worked with most began to locate for themselves some of the major images under which they had laboured during their life: images of the rest of their family, images of how you please your father and mother, how to be attractive to them, images of brothers and sisters, images of how you are a man or a woman, and of how all of these became enmeshed in the highly seductive images Chris Brain insinuated into their picture of themselves. By entering into some of these images in my own imagination, by witnessing them in this way, and sometimes by telling what they made me feel, I think

I helped them reappropriate what was theirs from what had been foisted on them. Most of all though, it was creating an atmosphere in which it was all right for them to talk and talk and talk.

It also seemed to help them to assume less guilt and shame about what had happened, and to feel less angry about it. To have undergone the experience they did cannot help but be part of their lives. When I last saw them however, they felt they were managing to let go of some of the more obsessive emotions associated with NOS. They were trying to resume their own lives together and making plans for a future free from the unconscious imprisonment into which they had been manipulated for such a large part of their lives.

It has sometimes been said that those who become members of cults do so because of personal needs of their own, that cult membership indicates some sort of personality problem. While this certainly seems true of cult leaders, as far as my clients are concerned, I doubt this. Of course, personal needs, of the kind we all have, become bound up in a cult like NOS, but all it needed for them to become involved was to attend a meeting in a church, led by a particular and very charismatic person, at a certain place in Sheffield ten years ago, when they had scarcely left their adolescence behind them. To all intents and purposes it was an accident. They themselves always appeared to be very perceptive and capable individuals who found themselves in a traumatising situation; there was little sign of anything in them that might predispose them to bring the misfortune they suffered on themselves. They could hardly expect that the exciting visionary of their own generation they first met on that fateful occasion would be so able to ensnare them in his fantasy.

It is as well that NOS ended when it did. A few months later and many of its members might have been in America, houses and belongings sold to support their leader, in a place where the local therapeutic help they had received in Sheffield might not have been available, and where, to give the Anglican Church its due, the means of defrocking the priest it had ordained might not have existed so readily. To invoke a comparison used in conversation with this couple, it was as if they had been involved in a shipwreck, and, distressing and frightening as this was, fortunately it had occurred near dry land instead of far away at sea.

My account of this work can be seen from two perspectives. The first, and more important, from an immediate point of view, was the situation of the two people worked with, and I hope I have done their experience justice. The second, which has been touched on already, has been method. To approach the task of helping someone who has suffered a traumatic shock in their life is no easy matter. There is so much one does not know about another's life, of its substances and practices, or the metaphorical resonances that these may stir up in one's own. It is tempting to take refuge in the techniques that often constitute professional identity. So, one might attempt to capture the life of others in a concrete procedure, image-making, talking in a way familiar to us, in a psychodynamics, theory or system inculcated in our training – all of which should be no more than a scaffolding for later work. But though, if it is not refused outright, this might buy a limited compliance, it will almost certainly miss the complex essence of their experience – for them as well as us. Certainly my clients were exhausted by the 'techniques of being' intruded into their lives by NOS. If there is a method that is useful, then it is a kind of systematic absent-mindedness, an open-focused attention, that assists those we are working with to light upon a style of imagery, gesture and narrative native to their concerns and predicaments. This style is not a technical matter: it cannot be understood by 'scientific' psychological analysis, quantifiable measurement, concentration on words, images, or any other expressive form enlisted in the interests of therapy taken in isolation. Rather it is akin to what Cox and Theilgaard call the 'Aeolian Mode' in psychotherapy, which '... facilitates response to the numerous nuances, and the hints of "other things" which so often people therapeutic space' (1987: xxvi).

To attend, see or listen closely, which is our part in a therapeutic engagement, cuts across the official modes of one or other psychotherapy or arts therapy, each of which, for complicated political reasons, may attempt to promote its own symbolic and expressive specialities. At root, these consist of hidden metaphorical or aesthetic preferences – the talking cure, the couch, Oedipus, the transference, the archetypes, the relationship between mother and child, image, theatre, music, and so on. Yet all of them may mediate personal and social relationships in powerful ways. Writing about what he calls the 'cross transference' between different sense perceptions, feelings and forms of thought in understanding art, but implicitly about all human acts, Nelson Goodman says: 'The absurd and awkward myth of the insularity of aesthetic experience can be scrapped' (1981: 260). (Of course the same might be said of scientific 'insularity'!) To make 'evidence based' claims for one of these therapeutic metaphors by itself is a damaging mistake, for how in any coherent sense can they be treated as if they were medical treatments, except to satisfy extraneous institutional and economic demands? If we can escape from such restrictive practice and be open and eclectic in our method, this will help us avoid being too concerned with practical utility, will be an antidote to sentimental solutions, and rescue us from banality; all three of which conspire as much against art as they do against psychotherapy and human understanding.

Note

1 In the same year that the conference at which this paper was given took place, the events with which it is concerned received much media attention, and a book was written that deals with them in detail (Howard 1996). While in writing this chapter, I have textually revised the original paper and made a few alterations, it remains largely as originally given. It arises from therapeutic support I offered members of the Sheffield Nine O'clock Service, following its collapse amidst allegations of sexual abuse levelled at its charismatic leader.

References

Bachelard, G. (1969) *The Poetics of Space*. Boston: Beacon Press.

Cox, M. and Theilgaard, A. (1987) *Mutative Metaphors in Psychotherapy: The Aeolian Mode*. London and New York: Tavistock Publications.

Goodman, N. (1981) *Languages of Art: An Approach to a Theory of Symbols*. Brighton: The Harvester Press.

Howard, R. (1996) *The Rise and Fall of the Nine O'clock Service: A Cult within the Church?* London and New York: Mowbray.

Index

Index compiled by
Auriol Griffith-Jones

Also published by FAB

ON ART AND THERAPY - An Exploration

Martina Thomson

'All through the book there are sentences and images that made me want to throw my hat in the air ... Don't underestimate it. The tone is modest, but it is honest.' **Seamus Heaney**

Here Martina Thomson makes an eloquent plea for the therapist to return to a trust in the therapeutic value of the creative process. Drawing on her memories of working with some of the great pioneers of art therapy, and on her own experience as a painter and as an art therapy practitioner, her argument surfaces gently through observation, speculation, case history, and quotations from artists, poets and analysts.

THE HEALING DRAMA

Psychodrama and Dramatherapy with Abused Children

Anne Bannister

Working with children who have been physically or sexually abused presents tremendous challenges. Various therapeutic techniques can be used, and one of them is through the medium of drama in the hands of psychodramatists and dramatherapists. This book sets out the main theoretical and practical approaches. Although specifically directed to professionals working in psychodrama, it will also be required reading for those in social work, child and educational psychology, teaching and play therapy.

THE LONE TWIN

Understanding Twin Bereavement and Loss

Joan Woodward

What happens when twins are separated, especially by death? Using Bowlby's Attachment Theory as her conceptual base, the author takes the reader through the closeness of being a twin, including its negative aspects and their need to be different from each other, illustrating the book with the words of lone twins themselves. An important and rare book for any professional working in bereavement counselling.

ACTING-IN - Practical Applications of Psychodrama Methods

Dr Adam Blatner

Since its first publication twenty-three years ago, *Acting-In* has now become one of the major introductory texts in the field of psychodrama and a significant resource for psychotherapists and others who use role playing or action techniques in their work. This edition has been significantly revised. Generously illustrated throughout with case examples, *Acting-In* presents the essentials of psychodrama and how to use them.

New Titles for 1998

THE BUTTERFLY AND THE SERPENT

Essays in Psychiatry, Race and Religion

Roland Littlewood

In the last twenty years cultural psychiatry and medical anthropology have become firmly established as academic and clinical disciplines in the United Kingdom and North America. Through his research and his writings, Roland Littlewood has emerged as one of the leading and most distinguished contributors to the field of cultural psychiatry. Here he presents a collection of his papers which have achieved great influence and which deal with the essential issues in this area. They range across culture, history, language, religion, and gender and present material from several continents. All of them provide clear theoretical positions through accessible narrative accounts.

For students, teachers and researchers in psychiatry, medicine, social anthropology, ethnic and racial studies, and medical sociology, this provides an outstanding and original overview of this fascinating and rapidly developing area.

JOHN BOWLBY: HIS EARLY LIFE

A Biographical Journey into the Roots of Attachment Theory

Suzan van Dijken

In an insightful treatment of the early years of John Bowlby's life and work, Suzan van Dijken sheds light on a number of events that are very much linked to the eventual creation of Attachment Theory but have not been known about or published to date. In addition, she provides much new information about topics that Bowlby was quite reluctant to discuss in detail, whether in public or in private, and yet are clearly connected to his later life and theoretical pursuits.

This biographical portrait covers in great depth Bowlby's family of origin, his upbringing, schooling and later education, his little-known work with Cyril Burt, his introduction to psychoanalysis, and his involvement in some of the major events in that world, including the Controversial Discussions between Anna Freud and Melanie Klein.

New Titles for 1998

MIDWIFERY OF THE SOUL

A Holistic Perspective on Psychoanalysis

Margaret Arden

Early on in her career, Margaret Arden became fascinated by the problem of the gap between theory and practice. She came to realise that what she relied on in her work was not the theory she had been taught but her sense of the truth of what was going on in the consulting room. The process of psychoanalysis, the work of transference and counter transference, enables the analyst to meet one of the patient's basic needs by recognising the truth of who he or she is.

This book takes the reader on an intellectual journey. Each chapter represents a stage in the exploration of ideas which have influenced the author's view of psychoanalysis. The book is remarkable for the coherance of her thinking and the clarity of presentation of ideas which connect Jung with Freud, science with religion and the emerging science of consciousness with Goethe's scientific views. Anyone who has felt the need to explore the unrealised possibility of psychoanalytic theory will find this book rewarding.

THE PSYCHOANALYTIC MYSTIC

Michael Eigen

Most psychoanalysts tend to be anti-mystical or, at least, non-mystical. Psychoanalysis is allied with science and, if anything, is capable of deconstructing mystical experience. Yet some psychoanalysts tend to be mystical or make use of the mystical experience as an intuitive model for psychoanalysis. Indeed, the greatest split in the psychodynamic movement, between Freud and Jung, partly hinged on the way in which mystical experience was to be understood.

Michael Eigen has often advocated and encouraged a return to the spiritual in psychoanalysis - what Freud called the 'oceanic feeling'. Here he expands on his call to celebrate and explore the meaning of mystical experience within psychoanalysis, illustrating his writing with the work of Bion, Milner and Winnicott.

New Titles for 1998

NEW PERSPECTIVES ON PSYCHOTHERAPY AND HOMOSEXUALITIES

Edited by Christopher Shelley

Most psychotherapy training programmes don't incorporate elements which examine the special needs of the gay and lesbian populations. It is therefore questionable whether practitioners possess the basic necessary skills for assessing and employing interventions based on sexually sensitive material. An unexamined and untrained approach to working with homosexual populations can no longer be tolerated.
This book address some of the incoherence that exists in this field. The contributors address theories and models of practice that will be more beneficial to the therapeutic needs of homosexual clients. Their accounts represent a significant step towards a better understanding of the needs of this client group.

THE ELUSIVE HUMAN SUBJECT

A Psychoanalytic Theory of Subject Relations
Roger Kennedy

'This is a major work on the nature of human subjectivity. Kennedy argues that we need to conceptualize the human subject differently in psychoanalysis, and he proposes a complex yet lucid theory of what that would be. I have no doubt that it will establish itself as one of the major works of contemporary psychoanalysis.' **Christopher Bollas**

'There is something essentially elusive about our subjective life, which makes it difficult to capture'. From this position of uncertainty, Kennedy pursues his exploration of how we can gain access to the human subject, through what we experience as individuals and also through the multiple and complex interactions between individuals in the social field.

This is an outstanding book in which Roger Kennedy has succeeded in a major re-evaluation of how we describe the Self, and what that means for psychoanalysis.

Also published by FAB

FREELY ASSOCIATED

Encounters in Psychoanalysis with Christopher Bollas, Nina Coltart, Michael Eigen, Joyce McDougall, Adam Phillips
Edited by Anthony Molino

From a series of interviews conducted over two years, this book offers a rewarding, fascinating and rare opportunity to encounter five extraordinary psychoanalysts speaking for themselves.

H.J.S. GUNTRIP - A Psychoanalytical Biography

Jeremy Hazell

'This extraordinary book does more than any ordinary biography could to rescue from the archives of the Menninger Clinic a uniquely intimate study of Harry Guntrip - the man described by John D. Sutherland as "one of the psychoanalytic immortals".
No analyst or therapist, from the newest student to the eminences grises, can fail to be absorbed by the detailed session-by-session story of two of the great figures of psycho-analysis at work in their consulting rooms. We see the theories, techniques but, more importantly, the personalities of these two Object-Relations pioneers evolving before us as we read. I was aware of a sense of privilege at being admitted to a unique event, the unfolding of the psychological history of a rare and engaging man.' **Nina Coltart**

AN INTRODUCTION TO OBJECT RELATIONS

Lavinia Gomez

'..enjoyable, stimulating and informative. I would recommend it to students of psychotherapy and counselling, and also to psychiatrists in training or in continuing professional development.'

British Journal of Psychiatry

In this critical introduction to the subject, Lavinia Gomez presents the work of the main theorists chronologically, enabling the reader to gain a sense of how Object Relations developed, and the way in which the theorists built on, diverged from and opposed each other's ideas. A brief biography brings to life the persons behind the theory, contributing to a deeper understanding and critical appreciation of their ideas.